Barron's Regents Exams and Answers

Geometry

LAWRENCE S. LEFF, M.S.
Former Assistant Principal
Mathematics Supervisor
Franklin D. Roosevelt High School, Brooklyn, NY

BARRON'S

Barron's Educational Series, Inc.

All inquiries should be addressed to:
Barron's Educational Series, Inc.
250 Wireless Boulevard
Hauppauge, NY 11788
www.barronseduc.com

ISBN: 978-0-7641-4222-2
ISSN: 1944-1959

PRINTED IN THE UNITED STATES OF AMERICA
9 8 7 6 5 4 3 2 1

10%
POST-CONSUMER
WASTE
Paper contains a minimum
of 10% post-consumer
waste (PCW). Paper used
in this book was derived
from certified, sustainable
forestlands.

Contents

Preface   **v**

How to Use This Book   **1**

Getting Acquainted with the Geometry Regents Exam   **4**

Ten Test-Taking Tips   **9**

1. Know What to Expect on Test Day
2. Avoid Last-Minute Studying
3. Be Well Rested and Come Prepared on Test Day
4. Know How and When to Use Your Calculator
5. Have a Plan for Budgeting Your Time
6. Make Your Answers Easy to Read
7. Answer the Question That Is Asked
8. Take Advantage of Multiple-Choice Questions
9. Don't Omit Any Questions
10. Mark Up Diagrams

A Brief Review of Key Geometry Facts and Skills 21

1. Angle and Parallel Line Relationships
2. Relationships in Triangles and Polygons
3. Parallelograms and Trapezoids
4. Special Right Triangle Relationships
5. Locus and Concurrency Theorems
6. Circles and Angle Measurement
7. Coordinate Relationships
8. Volume and Surface Area
9. Transformation Geometry
10. Logical Reasoning and Proof

Geometric Constructions 147

Basic Constructions
Required Constructions

Glossary of Terms 155

Regents Examinations, Answers, and Self-Analysis Charts 165

August 2009 Exam 167
June 2010 Exam 219
August 2010 Exam 274
June 2011 Exam 327
August 2011 Exam 381
June 2012 Exam 434

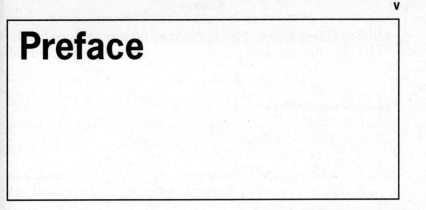

Preface

This book is designed to prepare you for the Geometry Regents examination while strengthening your understanding and mastery of the topics on which this test is based. In addition to providing questions from previous mathematics Regents examinations, this book offers these special features:

• **Step-by-Step Explanations of the Solutions to All Regents Questions**. Careful study of the solutions and explanations will improve your mastery of the subject. Each explanation is designed to show you how to apply the facts and concepts you have learned in class. Since the explanation for each solution has been written with emphasis on the reasoning behind each step, its value goes well beyond the application to a particular question.

• **Unique System of Self-Analysis Charts**. Each set of solutions for a particular Regents exam ends with a Self-Analysis Chart. These charts will help you to identify weaknesses and direct your study efforts where needed. In addition, the charts classify the questions on each exam into an organized set of topic groups. This feature will enable you to locate other questions on the same topic in other Geometry Regents exams.

- *General Test-Taking Tips*. Tips are given that will help to raise your grade on the actual Geometry Regents exam that you will take.

- *Geometry Refresher*. A brief review of key mathematics facts and skills that are tested on the Geometry Regents exam is included for easy reference and quick study.

- *Required Geometric Constructions*. Compass constructions that you are required to know are collected and explained in a separate section.

- *Glossary*. Definitions of important terms related to the Geometry Regents exam are conveniently organized in a glossary.

	Number of Questions					
Topic	August 2009	June 2010	August 2010	June 2011	August 2011	June 2012
1. Logic (negation, conjunction, disjunction, related conditionals, biconditional, logically equivalent statements, logical inference)	1	2	1	1	1	1
2. Angle & Line Relationships (vertical angles, midpoint, altitude, median, bisector, supplementary angles, complete angles, parallel and perpendicular lines)	2	1	2	1	1	2
3. Parallel & Perpendicular Planes	2	1	1	2	2	2
4. Angles of a Triangle & Polygon (sum of angles, exterior angle, base angles theorem, angles of a regular polygon)	2	2	3	3	4	3
5. Triangle Inequalities (side length restrictions, unequal angles opposite unequal sides, exterior angle)	2	1	1	—	—	—
6. Trapezoids & Parallelograms (includes properties of rectangle, rhombus, square, and isosceles trapezoid)	4	2	1	3	2	3
7. Proofs Involving Congruent Triangles	2	1	2	2	2	1
8. Indirect Proof & Mathematical Reasoning	—	—	—	—	—	—
9. Ratio & Proportion (includes similar polygons, proportions, and proofs involving similar triangles, comparing linear dimensions, and areas of similar triangles)	—	—	3	2	3	2
10. Proportions formed by altitude to hypotenuse of right triangle; Pythagorean theorem	1	1	1	1	2	—
11. Coordinate Geometry (slopes of parallel and perpendicular lines; slope-intercept and point-slope equations of a line; applications requiring midpoint, distance, or slope formulas; coordinate proofs)	6	6	6	6	7	6

	Number of Questions					
Topic	**August 2009**	**June 2010**	**August 2010**	**June 2011**	**August 2011**	**June 2012**
12. Solving a Linear-Quadratic System of Equalities Graphically	1	1	1	1	1	1
13. Transformation Geometry (includes isometries, dilation, symmetry, composite transformations, transformations using coordinates)	4	4	3	4	4	4
14. Locus & Constructions (simple and compound locus, locus using coordinates, compass constructions)	2	3	3	4	3	3
15. Midpoint and Concurrency Theorems (includes joining midpoints of two sides of a triangle, three sides of a triangle, the sides of a quadrilateral, median of a trapezoid; centroid, orthocenter, incenter, and circumcenter)	2	2	3	1	—	2
16. Circles & Angle Measurement (incl. ≅ tangents from same exterior point, radius ⊥ tangent, tangent circles, common tangents, arcs and chords, diameter ⊥ chord, arc length, area of a sector, center-radius equation, applying transformations)	4	4	4	4	4	4
17. Circles and Similar Triangles (includes proofs, segments of intersecting chords, tangent and secant segments)	1	2	1	1	—	2
18. Area of Plane Figures (includes area of a regular polygon)	—	1	—	—	—	—
19. Measurement of Solids (volume and lateral area; surface area of a sphere; great circle; comparing similar solids)	2	4	2	2	2	2

How to Use This Book

This section explains how you can make the best use of your study time. As you work your way through this book, you will be following a carefully designed five-step study plan that will improve your understanding of the topics that the Geometry Regents examination tests while raising your exam grade.

Step 1. *Know What to Expect on Test Day.* Before the day of the test, you should be thoroughly familiar with the format, the scoring, and the special directions for the Geometry Regents exam. This knowledge will help you build confidence and prevent errors that may arise from misunderstanding the directions. The next section in this book, "Getting Acquainted with the Geometry Regents Exam," provides this important information.

Step 2. *Become "Testwise."* The section titled "Ten Test-Taking Tips" will alert you to easy things that you can do to become better prepared and to be more confident when you take the actual test.

Step 3. *Review Geometry Topics.* Since the Geometry Regents exam will test you on topics that you may not have studied recently, the section entitled "A Brief Review of Key Geometry Facts and Skills" provides a quick refresher of the major topics tested on the examination. This section also includes illustrative practice exercises with worked-out solutions.

Step 4. Review Geometry Constructions. A separate sectio[n] identifies and explains in step-by-step fashion the geometric con[-] structions that you are required to know.

Step 5. Take Practice Exams Under Exam Conditions. Thi[s] book contains actual Mathematics Regents exams with carefull[y] worked-out solutions for all of the questions. When you reach thi[s] part of the book, you should do these things:

- After you complete an exam, check the answer key for the entir[e] test. Circle any omitted questions or questions that you answere[d] incorrectly. Study the explained solutions for these questions.

- On the Self-Analysis Chart, find the topic under which eac[h] question is classified and enter the number of points you earne[d] if you answered the question correctly.

- Figure out your percentage on each topic by dividing you[r] earned points by the total number of points allotted to that topic[,] carrying the division to two decimal places. If you are not satis[-] fied with your percentage on any topic, reread the explaine[d] solutions for the questions you missed. Then locate related ques[-] tions in other Regents examinations by using their Self-Analysi[s] Charts to see which questions are listed for the troublesom[e] topic. Attempting to solve these questions and then studyin[g] their solutions will provide you with additional preparation. Yo[u] may also find it helpful to review the appropriate sections in "[A] Brief Review of Key Geometry Facts and Skills." More detaile[d] explanations and additional practice problems with answers ar[e] available in Barron's companion book, *Let's Review: Geometry.*

Tips for Practicing Effectively and Efficiently

- When taking a practice test, do not spend too much time on any one question. If you cannot come up with a method to use, or if you cannot complete the solution, put a slash through the number of the question. When you have completed as many questions as you can, return to the unanswered questions and try them again.

- After finishing a practice test, compare each of your solutions with the solutions that are given. Read the explanation provided even if you have answered the question correctly. Each solution has been carefully designed to provide additional insight that may be valuable when answering a more difficult question on the same topic.

- In the weeks before the actual test, plan to devote at least ½ hour each day to preparation. It is better to spread out your time in this way than to cram by preparing for, say, 4 hours the day before the exam will be given. As the test day gets closer, take at least one complete Regents exam under actual test conditions.

Getting Acquainted with the Geometry Regents Exam

This section explains things about the Geometry Regents examination that you may not know, such as how the exam is organized, how your exam will be scored, and where you can find a complete listing of the topics tested by this exam.

WHEN DO I TAKE THE GEOMETRY REGENTS EXAM?

The Geometry Regents exam is administered in January, June, and August of every school year. Most students will take this exam after successfully completing a 1-year high-school-level geometry course.

HOW IS THE GEOMETRY REGENTS EXAM SET UP?

The Geometry Regents exam is divided into four parts with a total of 38 questions. All of the questions in each of the four parts must be answered. You will be allowed a maximum of 3 hours in which to complete the test.

- Part I consists of 28 standard multiple-choice questions, each with four answer choices labeled (1), (2), (3), and (4).

- Parts II, III, and IV contain six, three, and one question(s), respectively. The answers and the accompanying work for the questions in these three parts must be written directly in the question booklet. You must show or explain how you arrived at each answer by indicating the necessary steps, including appropriate formula substitutions, diagrams, graphs, and charts. If you use a guess-and-check strategy to arrive at a numerical answer for a problem, you must indicate your method and show the work for at least three guesses.

- Since scrap paper is not permitted for any part of the exam, you may use the blank spaces in the question booklet as scrap paper. If you need to draw a graph, graph paper will be provided in the question booklet. All work should be done in pen, except graphs and diagrams, which should be drawn in pencil.

The accompanying table summarizes how the Geometry Regents exam breaks down.

Question Type	Number of Questions	Credit Value
Part I: Multiple choice	28	$28 \times 2 = 56$
Part II: 2-credit open-ended	6	$6 \times 2 = 12$
Part III: 4-credit open-ended	3	$3 \times 4 = 12$
Part IV: 6-credit open-ended	1	$1 \times 6 = 6$
	Total = 38 questions	Total = 86 points

WHAT TYPE OF CALCULATOR DO I NEED?

Graphing calculators are *required* for the Geometry Regents examination. During the administration of the Regents exam, schools are required to make a graphing calculator available for the exclusive use of each student. You will need to use your calculator to work with trigonometric functions of angles, evaluate roots, and perform routine calculations. Knowing how to use a graphing calculator gives you an advantage when deciding how to solve a problem. Rather than solving a problem algebraically with pen and paper, it may be easier to solve the same problem using a graph or table created by a graphing calculator. A graphical or numerical solution using a calculator can also be used to help confirm an answer obtained using standard algebraic methods.

WHAT IS COLLECTED AT THE END OF THE EXAMINATION?

At the end of examination, you must return:

- Any permitted tool provided to you by your school such as a graphing calculator.

- The question booklet. Check that you have printed your name and the name of your school in the appropriate boxes near the top of the first page.

- The Part I answer sheet, as well as any other special answer form that is provided by your school. You must sign the statement at the bottom of the Part I answer sheet indicating that you have not received any unlawful assistance in answering any of the questions. If you fail to sign this declaration, your answer paper will not be accepted.

HOW IS THE EXAM SCORED?

Your answers to the 28 multiple-choice questions in Part I are scored as either correct or incorrect. Each correct answer receives 2 points. The six questions in Part II are worth 2 points each, the three questions in Part III are worth 4 points each, and the question in Part IV is worth 6 points. Solutions to questions in Parts II, III, and IV that are not completely correct may receive partial credit according to a special scoring guide provided by the New York State Education Department.

HOW IS YOUR FINAL SCORE DETERMINED?

The maximum total raw score for the Geometry Regents examination is 86 points. After the raw scores for the four parts of the test are added together, a conversion table provided by the New York State Education Department is used to convert your raw score into a final test score that falls within the usual 0 to 100 scale.

ARE ANY FORMULAS PROVIDED?

The Geometry Regents Examination test booklet will include a reference sheet containing the formulas included in the following series of boxes. This formula sheet, however, does not necessarily include all of the formulas that you are expected to know.

	Cylinder	$V = Bh,$ where B is the area of the base
Volume	Pyramid	$V = \dfrac{1}{3}Bh,$ where B is the area of the base
	Right circular cone	$V = \dfrac{1}{3}Bh,$ where B is the area of the base
	Sphere	$V = \dfrac{4}{3}\pi r^3$

Lateral area (L)	Right circular cylinder	$L = 2\pi rh$
	Right circular cone	$L = \pi r\ell$, where ℓ is the slant height

Surface area	Sphere	$SA = 4\pi r^2$

WHAT IS THE CORE CURRICULUM?

The *Core Curriculum* is the official publication by the New York State Education Department that describes the topics and skills required by the Geometry Regents examination. The *Core Curriculum* for Geometry includes most of the topics previously included in the geometry units for Math A *and* Math B. It also includes some additional geometry topics such as midpoint and concurrency theorems, similarity theorems, logical connectives, and aspects of solid geometry including parallel and perpendicular planes. If you have Internet access, you can view the *Core Curriculum* at the New York State Education Department's web site at *http://www.emsc.nysed.gov/3-8/MathCore.pdf*

Ten Test-Taking Tips

1. Know What to Expect on Test Day
2. Avoid Last-Minute Studying
3. Be Well Rested and Come Prepared on Test Day
4. Know How and When to Use Your Calculator
5. Have a Plan for Budgeting Your Time
6. Make Your Answers Easy to Read
7. Answer the Question That Is Asked
8. Take Advantage of Multiple-Choice Questions
9. Don't Omit Any Questions
10. Mark Up Diagrams

These ten practical tips can help you raise your grade on the Geometry Regents examination.

TIP 1

Know What to Expect on Test Day

SUGGESTIONS
- Become familiar with the format, directions, and content of the Geometry Regents exam.

- Know where you should write your answers for the different parts of the test.

- Ask your teacher to show you an actual test booklet from a previously given Math Regents exam.

TIP 2

Avoid Last-Minute Studying

SUGGESTIONS

- Start your Geometry Regents examination preparation early by making a regular practice of:

  1. taking detailed notes in class and then reviewing your notes when you get home;
  2. completing all written homework assignments in a neat and organized way;
  3. writing down any questions you have about your homework so that you can ask your teacher about them; and
  4. saving your classroom tests for use as an additional source of practice questions.

- Get a review book early in your preparation so that additional practice examples and explanations, if needed, will be at your fingertips. The recommended review book is Barron's *Let's Review: Geometry*. This easy-to-follow book has been designed for fast and effective learning and includes numerous demonstration and practice examples with solutions as well as graphing calculator approaches.

- Build your skill and confidence by completing all of the exams in this book and studying the accompanying solutions before the day of the Geometry Regents exam. Because each exam takes up to 3 hours to complete, you should begin this process no later than several weeks before the exam is scheduled to be given.

- As the day of the actual exam nears, take the exams in this book under the timed conditions that you will encounter on the actual test. Then compare each of your answers with the explained answers given in this book.

- Use the Self-Analysis Chart at the end of each exam to help pinpoint any weaknesses.

- If you do not feel confident in a particular area, study the corresponding topic in Barron's *Let's Review: Geometry*.

- As you work your way through the exams in this book, make a list of any formulas or rules that you need to know and learn them well before the day of the exam. The formulas that will be provided in the test booklet are listed on pages 7–8.

TIP 3

Be Well Rested and Come Prepared on Test Day

SUGGESTIONS

- On the night before the Regents exam, lay out all of the things you must take with you. Check the items against the following list:

  - ☐ Your exam room number as well as any personal identification that your school may require.
  - ☐ Two pens.
  - ☐ Two sharpened pencils with erasers.
  - ☐ A ruler and a compass.
  - ☐ A graphing calculator (with fresh batteries).
  - ☐ A watch.

- Eat wisely and go to bed early so you will be alert and well rested when you take the exam.

- Be certain you know when your exam begins. Set your alarm clock to give yourself plenty of time to eat breakfast and travel to school. Also, tell your parents what time you will need to leave the house in order to get to school on time.

- Arrive at the exam room on time and with confidence that you are well prepared.

TIP 4

Know How and When to Use Your Calculator

SUGGESTIONS

- Bring to the Geometry Regents exam room the same calculator that you used when you completed the practice exams at home. Although a graphing calculator is required for the Geometry Regents exam, don't expect to have to use your calculator on each question. Most questions will not require a calculator.

- Avoid careless errors by using your calculator to perform routine arithmetic calculations that are not easily performed mentally.

- If you are required to use a graphing calculator provided by your school, make sure that you practice with it in advance because not all calculators work in the same way.

- Before you begin the test, check that the calculator you are using is set to degree mode.

- Avoid rounding errors. Unless otherwise directed, the π (pi) key on a calculator should be used in computations involving the constant π rather than the familiar rational approximation of 3.14 or $\frac{22}{7}$. When performing a sequence of calculations in which the result of one calculation is used in a second calculation, do not round off. Instead, use the full power/display of the calculator by saving intermediate results in the calculator's memory.

Unless otherwise directed, rounding, if required, should be done only when the *final* answer is reached.

- Because it is easy to press the wrong key, first estimate an answer and then compare it to the answer obtained by using your calculator. If the two answers are very different, start over.

TIP 5

Have a Plan for Budgeting Your Time

SUGGESTIONS

- In the **first 90 minutes** of the 3-hour Geometry Regents exam, complete the 28 multiple-choice questions in Part I. When answering troublesome questions of this type, first rule out any choices that are impossible. If the choices are numbers, you may be able to identify the correct answer by plugging these numbers back into the original question to see which one works.

- During the **next 45 minutes** of the exam, complete the six Part II questions and the three Part III questions. To maximize your credit for each question, write down clearly the steps you followed to arrive at each answer. Include any equations, formula substitutions, diagrams, tables, graphs, and so forth.

- In the **next 20 minutes** of the exam, complete the Part IV question. If a proof is required, be sure to explain your work in a clear, step-by-step fashion using valid methods of reasoning.

- During the **remaining 25 minutes**, review your entire test paper for neatness, accuracy, and completeness.

  1. Check that all the answers (except graphs and other drawings) are written in ink. Make sure you have answered all of the questions in each part of the exam and that all of your Part I answers have been transcribed accurately on the separate Part I answer sheet.

2. Before you submit your test materials to the proctor, check that you have written your name in the reserved spaces on the front page of the question booklet and on the Part I answer sheet. Also, don't forget to sign the declaration that appears at the bottom of the Part I answer sheet.

TIP 6

Make Your Answers Easy to Read

SUGGESTIONS

- Make sure your solutions and answers are clear, neat, and logically organized. When solving problems algebraically, define what the variables stand for, as in "Let x = ..."

- Use a pencil to draw graphs so that you can erase neatly, if necessary. Use a ruler to draw straight lines and coordinate axes.

- Label the coordinate axes. Write y at the top of the vertical axis, and write x to the right of the horizontal axis. Label each graph with its equation.

- When answering a question in Part II, III, or IV, indicate your approach as well as your final answer. If a formula is involved, write the formula and then show the numerical substitutions. When writing an informal proof, provide enough details to enable someone who doesn't know how you think to understand why and how you moved from one step of the proof to the next and why your reasoning is valid. If the teacher who grades your paper finds it difficult to figure out what you have written, he or she may simply decide to mark your work as incorrect and to give you little, if any, partial credit.

> ### TIP 7
>
> ---
>
> **Answer the Question That Is Asked**

SUGGESTIONS

- Make sure each of your answers is in the form required by the question. For example, if a question asks for an approximation, round off your answer to the required decimal position. If a question requires that the answer be expressed in lowest terms, make sure that the numerator and denominator of a fractional answer do not have any common factors other than 1 or -1. For example, instead of leaving an answer as $\dfrac{10x^2}{15x}$, write it as $\dfrac{2x}{3}$. If a question calls for the answer in simplest form, make sure you simplify a square root radical so that the radicand does not contain any perfect square factors greater than 1. For example, instead of leaving an answer as $\sqrt{18}$, write $3\sqrt{2}$.

- If a question asks only for the x-coordinate (or y-coordinate) of a point, do *not* give both the x-coordinate and the y-coordinate.

- After solving a word problem, check the original question to make sure your *final* answer is the quantity that the question asks you to find.

- If units of measurement are given, as in area problems, check that your answer is expressed in the correct units.

- Make sure the units used in a computation are consistent. For example, when finding the volume of a rectangular box, its three dimensions must be expressed in the same units of measurement. Suppose a rectangular box measures 2 feet in width, 6 feet in length, and 8 inches in height. To find the number of cubic *feet* in the volume of the box, first convert 8 inches to an equivalent number of feet. Since 8 inches is equivalent to $\frac{8}{12} = \frac{2}{3}$ ft, the volume of the box is 2 ft $\times$ 6 ft $\times \frac{2}{3}$ ft = 8 ft^3.

- If a question requires a positive root of a quadratic equation, as in geometric problems in which the variable usually represents a physical dimension, make sure you reject the negative root.

TIP 8

Take Advantage of Multiple-Choice Questions

SUGGESTIONS

- If you get stuck on a multiple-choice question that includes a diagram, you may be able to use the diagram to help exclude answer choices.

EXAMPLE

In the accompanying diagram of $\triangle ABC$, $\overline{DE}$ is parallel to $\overline{BC}$. If $AD = 3$, $DB = 6$, and $DE = 5$, what is the length of $\overline{BC}$?

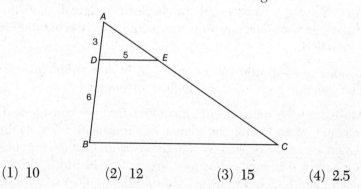

(1) 10   (2) 12   (3) 15   (4) 2.5

Solution 1: Unless otherwise indicated, figures that accompany questions are drawn approximately to scale. Since the length of

$\overline{BC}$ appears to be more than twice the length of $\overline{DE}$, you can eliminate choices (1) and (4). If you do not know how to solve the problem, then you can guess from among the remaining two choices.

Solution 2: Use the appropriate geometric relationship and solve algebraically. If a line that intersects two sides of a triangle is parallel to the third side, then it creates a triangle similar to the original triangle. Because $\triangle ADE \sim \triangle ABC$, lengths of corresponding sides are in proportion.

$$\frac{AD}{AB} = \frac{DE}{BC}$$

$$\frac{3}{3+6} = \frac{5}{BC}$$

$$3(BC) = 45$$

$$\frac{\cancel{3}(BC)}{\cancel{3}} = \frac{45}{3}$$

$$BC = 15$$

Hence, the correct choice is **(3)**.

TIP 9

Don't Omit Any Questions

SUGGESTIONS

- Keep in mind that on each of the four parts of the test you must answer all of the questions.

- If you get stuck on a multiple-choice question in Part I, try to eliminate any unlikely or impossible answers. Then guess.

- If you cannot fully answer a question from Part II, III, or IV, try to maximize your partial credit by writing any formula or mathematics facts that you think might apply. If appropriate, organize

and analyze the given information by making a table or a diagram. You may be able to arrive at the correct answer by guess-
ing, checking your guess, and then revising your guess, as
needed. Keep in mind that if a trial-and-error procedure is used,
you must show the work for at least three guesses.

• If you get stuck on a similarity or congruence proof, mark up the
diagram with the Given. If a diagram is not provided, draw your
own. Mark the diagram with any additional pairs of parts that you
can deduce as congruent such as: vertical angles; congruent
angles formed by perpendicular lines, parallel lines, or an angle
bisector; congruent segments formed by a midpoint, bisector, or
median. Write a plan that states the pair of triangles that must be
proved congruent (or similar), as well as the method to be used
(for example, SAS ≅ SAS). If you are required to prove a proportion or pairs of products of side lengths equal, reason backward
from the Prove to help identify the triangles that need to be
proved similar.

TIP 10

Mark Up Diagrams

SUGGESTIONS

• When a figure accompanies a question that requires some
numerical or algebraic calculation, organize the information by
labeling the diagram with any measures that are given or which
you are able to deduce from what is given. Also, label the
unknown quantity with a letter such as x. If no diagram is provided, draw your own.

• When a figure accompanies a proof, mark it up with any pair of
parts that are given as congruent, parallel, or perpendicular.
Also, mark off any other pairs of parts that you can conclude are
congruent.

- Do *not* assume segments or angles are congruent merely because they look that way in a diagram. Do *not* assume lines are parallel, are perpendicular, or are bisectors simply because they look so in a diagram. You can draw such conclusions only if they are included in the Given or when there is valid geometric justification.

EXAMPLE 1

The altitude drawn to the hypotenuse of a right triangle divides the hypotenuse into segments of lengths 4 cm and 12 cm. What is the number of centimeters in the length of the shorter leg of the right triangle?

Solution:
- Draw a diagram to help organize what you know about the problem. Represent the length of the shorter leg by x.

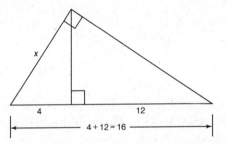

- Think of the relationship that involves an altitude drawn to the hypotenuse of a right triangle. When an altitude is drawn to the hypotenuse of a right triangle, each leg is the mean proportional between the segment of the hypotenuse adjacent to that leg and the whole hypotenuse.

- Using this relationship and the diagram, write an equation:

$$\frac{4}{x} = \frac{x}{16}$$

$$x^2 = 64$$

$$x = \sqrt{64} = 8$$

The length of the shorter leg is **8 cm**.

EXAMPLE 2

In the accompanying diagram, $\overline{AB}$ and $\overline{CD}$ intersect at E, E is the midpoint of $\overline{AB}$, and $\angle A \cong \angle B$.

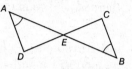

Which statement can be used to prove $\triangle ADE \cong \triangle BCE$?
(1) ASA $\cong$ ASA   (3) SSS $\cong$ SSS
(2) HL $\cong$ HL   (4) SAS $\cong$ SAS

Solution: Mark up the diagram with the Given. Also mark off the congruent vertical angles.

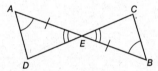

It should be clear from the diagram that the two triangles are congruent by ASA $\cong$ ASA. If you did not realize this, you can reason as follows:

- Since the Given includes the fact $\angle A \cong \angle B$, it is likely that the correct answer choice includes at least one pair of congruent angles. Therefore, you can eliminate choices (2) and (3).

- The diagram further shows that a second pair of angles are also congruent. Only answer choice (1), ASA $\cong$ ASA, includes two pairs of angles. Since the included sides, $\overline{AE}$ and $\overline{BE}$, are congruent from the Given, this is the correct method.

The correct answer is choice **(1)**.

A Brief Review of Key Geometry Facts and Skills

1. ANGLE AND PARALLEL LINE RELATIONSHIPS

1.1 PAIRS OF ANGLES
• Vertical angles are congruent:

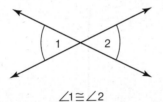

$$\angle 1 \cong \angle 2$$

• The measures of **supplementary angles** add up to 180, and the measures of **complementary angles** add up to 90.

1.2 PARALLEL AND PERPENDICULAR LINES
Parallel lines are lines in the same plane that do not intersect. If two parallel lines are cut by a transversal, any two of the eight angles that are formed are either congruent or supplementary. If in the accompanying figure $p \parallel q$, then

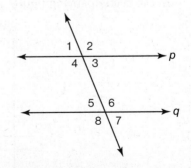

- Alternate interior angles are congruent:

$$\angle 3 \cong \angle 5 \text{ and } \angle 4 \cong \angle 6.$$

- Corresponding angles are congruent:

$$\angle 1 \cong \angle 5, \angle 4 \cong \angle 8, \angle 2 \cong \angle 6, \text{ and } \angle 3 \cong \angle 7.$$

- Alternate exterior angles are congruent:

$$\angle 1 \cong \angle 7 \text{ and } \angle 2 \cong \angle 8.$$

- Same side interior angles are supplementary:

$$m\angle 4 + m\angle 5 = 180 \text{ and } m\angle 3 + m\angle 6 = 180.$$

Perpendicular lines are lines that intersect at right angles.

- Two lines are perpendicular if they intersect to form congruent adjacent angles. In the accompanying figure, if $\angle 1 \cong \angle 2$, then $p \perp k$.

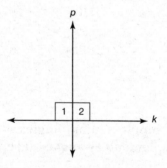

- Two lines are parallel if they are perpendicular to the same line. In the accompanying figure, if $p \perp k$ and $q \perp k$, then $p \parallel q$.

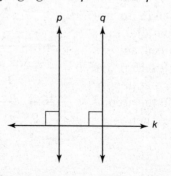

1.3 PARALLEL AND PERPENDICULAR PLANES

Consider the rectangular solid in the accompanying figure.

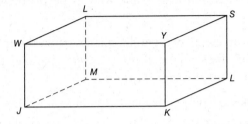

- If a line is perpendicular to each of two intersecting lines at their point of intersection, then the line is perpendicular to the plane determined by them. Line *JW* is perpendicular to lines *WL* and *WY*, so $\overline{JW} \perp$ plane *LWY*.

- Through a given point there passes one and only one plane perpendicular to a given line. Through point *L* only plane *LWY* is perpendicular to $\overleftrightarrow{LM}$.

- Through a given point there passes one and only one line perpendicular to a given plane. Through point *L* only $\overleftrightarrow{LM}$ is perpendicular to plane *LWY*.

- Two lines perpendicular to the same plane are coplanar. Lines *YW* and *SL* are each perpendicular to plane *WJM* and, as a result, are in the same plane (*LWY*).

- Two planes are perpendicular to each other if and only if one plane contains a line perpendicular to the second plane. Planes *LWJ* and *SLW* are perpendicular to each other if and only if line *LM* in plane *LWJ* is perpendicular to plane *SLW*.

- If a line is perpendicular to a plane, then any line perpendicular to the given line at its point of intersection with the given plane is in the given plane. Line *LM* is perpendicular to plane *MJK*, so $\overline{JM}$ is in plane *MJK*.

- If a line is perpendicular to a plane, then every plane containing the line is perpendicular to the given plane. Line *JW* is perpendicular to plane *LWY*, so plane *KJW* is perpendicular to plane *LWY*.

- If a plane intersects two parallel planes, then the intersection is two parallel lines. Plane *WJM* intersects parallel planes *LWY* and *MJK* at lines *WL* and *JM*, so $\overline{WL} \parallel \overline{JM}$.

- If two planes are perpendicular to the same line, they are parallel.

  Planes *WJM* and *YKL* are perpendicular to $\overrightarrow{SL}$ and are, therefore, parallel.

2. RELATIONSHIPS IN TRIANGLES AND POLYGONS

2.1 TRIANGLE RELATIONSHIPS

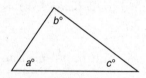

$$a + b + c = 180$$

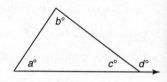

$$d = a + b$$

Isosceles Triangle

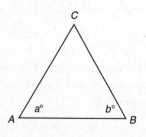

Equilateral Triangle

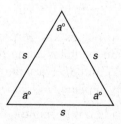

- If $\overline{AC} \cong \overline{BC}$, then $a = b$.

- If $a = b$, then $\overline{AC} \cong \overline{BC}$.

- $a = a = a = 60°$

- Area $= \dfrac{s^2 \sqrt{3}}{4}$

2.2 INEQUALITIES IN A TRIANGLE

Pairs of unequal angles of a triangle are opposite unequal sides with the larger angle facing the longer side.

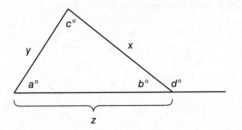

- If $x > y$, then $a > b$.
- If $a > b$, then $x > y$.
- $d > a$ and $d > c$.

The length of each side of a triangle is less than the sum of the lengths of the other two sides:

$$z < x + y, \; x < y + z, \text{ and } y < x + z.$$

Furthermore, each side of a triangle is longer than the difference in the lengths of the other two sides. For example, if the lengths of two sides of a triangle are 9 and 4, then the length of the third side, say x, must be shorter than $9 + 4 = 13$ and longer than $9 - 4 = 5$, as illustrated in the accompanying figure.

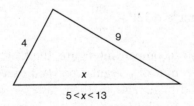

$5 < x < 13$

2.3 ANGLE RELATIONSHIPS IN POLYGONS

In any polygon with n sides:

- The sum of the measures of the n exterior angles, one at each corner, is 360. See the accompanying figure.

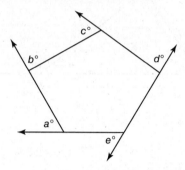

- The sum S of the interior angles is given by the formula $S = (n - 2) \times 180$. Thus, the four angles of a quadrilateral add up to $(4 - 2) \times 180 = 360°$; the five angles of a pentagon add up to $(5 - 2) \times 180 = 3 \times 180 = 540°$; and so forth.

A **regular polygon** is a polygon in which each angle has the same measure (equiangular) and each side has the same length (equilateral). In any regular polygon with n sides:

- Each exterior angle = $\dfrac{360}{n}$

- Each interior angle = $180 - \dfrac{360}{n}$

Referring to the accompanying figure, the length of the perpendicular from the center of a regular polygon to a side is called an **apothem**, denoted by a. The apothem also represents the radius of the circle that can be inscribed in a regular polygon. The apothem is the perpendicular bisector of the side to which it is drawn. If the perimeter of a regular polygon with n sides is represented by p, then:

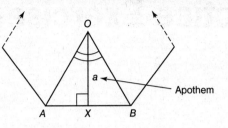

- Central $\angle AOB = \dfrac{360}{n}$

- $\overline{OX} \perp \overline{AB}$ and $\overline{AX} \cong \overline{BX}$

- Area $= \dfrac{1}{2} \times a \times p$

Practice Exercises

1. In the accompanying diagram, $\overrightarrow{AB}$ and $\overrightarrow{CD}$ intersect at E. If m$\angle AEC = 2x + 40$ and m$\angle CEB = x + 20$, find x.

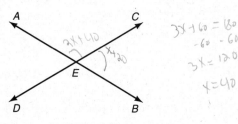

$$3x + 60 = 180$$
$$-60 \quad -60$$
$$3x = 120$$
$$x = 40$$

2. In the accompanying diagram, $\overline{AB} \perp \overline{BC}$, $\overline{DB} \perp \overline{BE}$, m$\angle CBE = x$, m$\angle DBC = y$, and m$\angle ABD = z$. Which statement *must* be true?

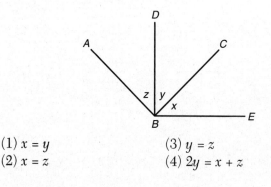

(1) $x = y$

(2) $x = z$

(3) $y = z$

(4) $2y = x + z$

3. The measures of the three angles of a triangle are in the ratio 2:3:4. Find the measure of the largest angle of the triangle.

4. In the accompanying diagram of $\triangle ABC$, $\overline{AB} \cong \overline{AC}$, $\overline{BD}$ and $\overline{DC}$ are angle bisectors, and m$\angle BAC$ = 20. Find m$\angle BDC$.

5. In the accompanying diagram of $\triangle ABC$, $\overline{BD}$ is drawn so that $\overline{BD} \cong \overline{DC}$. If m$\angle C$ = 70, find m$\angle BDA$.

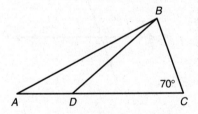

6. In $\triangle ABC$, side $\overline{AC}$ is extended through C to D. If m$\angle DCB$ = 50, which is the longest side of $\triangle ABC$?

7. If the degree measures of two complementary angles are in the ratio 2:13, what is the measure of the smaller angle?

8. In the accompanying diagram, $\overrightarrow{AB}$, $\overrightarrow{CD}$, $\overrightarrow{EF}$, and $\overrightarrow{GH}$ are straight lines. If m∠w = 30, m∠x = 30, and m∠z = 120, find m∠y.

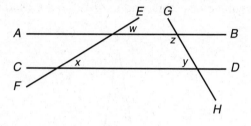

9. In the accompanying diagram, transversal $\overrightarrow{RS}$ intersects parallel lines $\overrightarrow{MN}$ and $\overrightarrow{PQ}$ at A and B, respectively. If m∠RAN = $3x + 24$ and m∠RBQ = $7x - 16$, find the value of x.

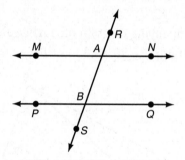

10. In the accompanying figure, $\overleftrightarrow{AB}$ is parallel to $\overleftrightarrow{CD}$, $\overline{AE} \perp \overline{BC}$, $m\angle BCD = 2x$, and $m\angle BAE = 3x$. Find the value of x.

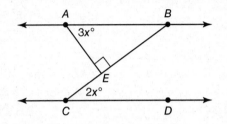

11. Find the number of sides in a regular polygon in which the measure of an interior angle is 144.

12. In the accompanying diagram, isosceles $\triangle ABC \cong$ isosceles $\triangle DEF$, $m\angle C = 5x$, and $m\angle D = 2x + 18$. Find $m\angle B$ and $m\angle BAG$.

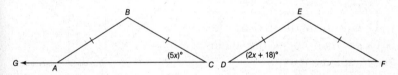

13. Which set *cannot* represent the lengths of the sides of a triangle?
(1) {4,5,6}   (3) {7,7,12}
(2) {5,5,11}   (4) {8,8,8}

14. The measures of five of the interior angles of a hexagon are 150, 100, 80, 165, and 150. What is the measure of the sixth interior angle?
(1) 75   (3) 105
(2) 80   (4) 180

Solutions

1. $\angle AEC$ and $\angle CEB$ are supplementary angles. The sum of the measures of two supplementary angles is 180:

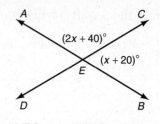

$$m\angle AEC + m\angle CEB = 180$$
$$2x + 40 + x + 20 = 180$$
$$3x + 60 = 180$$
$$3x = 180 - 60$$
$$= 120$$
$$x = 40$$

$x = \mathbf{40}$.

2. Perpendicular lines meet at right angles:

$\angle ABC$ is a right angle.

$\angle DBE$ is a right angle.

If the sum of two angles equals a right angle, the angles are complementary:

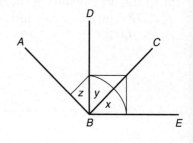

z is the complement of y,

x is the complement of y.

Complements of the same angle are equal in measure: $x = z$.

The correct choice is **(2)**.

3. Let $2x$, $3x$, and $4x$ represent, respectively, the degree measures of the three angles of the triangle.

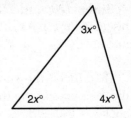

The sum of the measures of the three angles of a triangle is 180:

$$2x + 3x + 4x = 80$$
$$9x = 180$$

Divide both sides of the equation by 9:

$$x = 20$$

The measure of the largest angle is $4x$:

$$4x = 4(20) = 80$$

The measure of the largest angle is **80**.

4. The sum of the measures of the three angles of a triangle is 180:

$m\angle A = 20$:

$$m\angle A + m\angle B = 100$$
$$20 + m\angle B + m\angle C = 180$$
$$m\angle B + m\angle C = 180 - 20$$
$$= 160$$

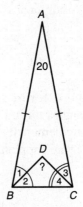

Since $\overline{AB} \cong \overline{AC}$, $\triangle ABC$ is isosceles; the base angles of an isosceles triangle are equal in measure.

$$m\angle B = m\angle C$$
$$2(m\angle B) = 160$$
$$m\angle B = 80$$
$$m\angle C = 80$$

Since $\overline{DB}$ and $\overline{DC}$ are angle bisectors, m∠1 = m∠2 and m∠3 = m∠4:

$$2(m\angle 2) = 80 \quad 2(m\angle 4) = 80$$
$$m\angle 2 = 40 \qquad m\angle 4 = 40$$

In △BDC, the sum of the measures of the three angles is 180:

$$m\angle BDC \;+ m\angle 2 + m\angle 4 = 180$$
$$m\angle BDC \;+ 40 \;+\; 40 \;= 180$$
$$m\angle BDC \;+\; 80 \;= 180$$
$$m\angle BDC = 180 - 80$$
$$= 100$$

m∠BDC = **100**.

5. Since $\overline{BD} \cong \overline{DC}$, △BCD is isosceles. The base angles of an isosceles triangle are equal in measure:

$$m\angle DBC = m\angle C = 70$$

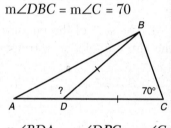

∠ADB is an exterior angle of △BDC. The measure of an exterior angle of a triangle is equal to the sum of the measures of the two remote interior angles:

$$m\angle BDA = m\angle DBC + m\angle C$$
$$m\angle BDA = \;\;70 \;\;\;+ 70$$
$$= \;\;140$$

m∠BDA = **140**.

6. If the exterior sides of two adjacent angles form a straight line, they are supplementary: thus ∠ACB and ∠BCD are supplementary.

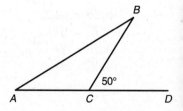

The sum of the measures of two supplementary angles is 180:

$$m\angle ACB + m\angle BCD = 180$$
$$m\angle ACB + 50 = 180$$
$$m\angle ACB = 180 - 50$$
$$ = 130$$

There can be only one obtuse angle in a triangle. Since $\angle ACB$ is an obtuse angle, it is the largest angle in $\triangle ABC$.

In a triangle, the side opposite the largest angle is the longest side.

Hence, $\overline{AB}$ is the longest side of $\triangle ABC$. (This side can also be designated as **AB** or **C**.)

7. Let $2x$ = degree measure of the smaller angle. Then $13x$ = degree measure of the larger angle.

Since the two angles are complementary, the sum of their degree measures is 90. Thus:

$$2x + 13x = 90$$
$$15x = 90$$
$$x = \frac{90}{15} = 6$$
$$2x = 2(6) = 12$$

The measure of the smaller angle is **12**.

8.

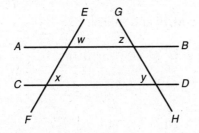

Since $m\angle w = 30$ and $m\angle x = 30$: $m\angle w = m\angle x$

If a transversal to two lines makes a pair of corresponding angles equal in measure, the lines are parallel: $\overline{AB} \parallel \overline{CD}$

If two lines are parallel, the interior angles on the same side of a transversal are supplementary, so $\angle y$ is the supplement of $\angle z$.

The sum of the measures of two supplementary angles is 180:

It is given that $m\angle z = 120$:

$$m\angle y + m\angle z = 180$$
$$m\angle y + \quad 120 = 180$$
$$m\angle y = 180 - 120$$
$$= 60$$

$m\angle y = \mathbf{60}$.

9.

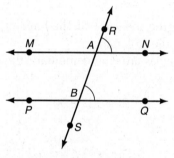

Angles RBQ and RAN are corresponding angles. If two lines are parallel, then corresponding angles have the same measure:

On each side, add 16 and subtract $3x$:

Divide each side of the resulting equation by 4:

$$m\angle RBQ = m\angle RAN$$
$$7x - 16 = \quad 3x + 24$$
$$\underline{-3x + 16 = -3x + 16}$$
$$4x \quad = \quad\quad 40$$

$$\frac{4x}{4} = \frac{40}{4}$$
$$x = 10$$

The value of x is **10**.

10. Since alternate interior angles formed by parallel lines have the same degree measure, m$\angle ABE$ = m$\angle BCD$ = $2x$. In right triangle AEB, angles BAE and ABE are the acute angles and, as a result, are complementary. Hence:

m$\angle BAE$ + m$\angle ABE$ = 90

$3x + 2x = 90$

$5x = 90$

$x = \dfrac{90}{5}$

$= 18$

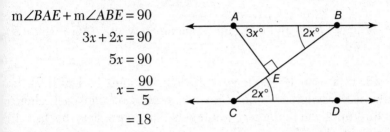

11. At each of the vertices of a regular polygon, the interior and exterior angles are supplementary.

- If the measure of an interior angle is 144, then the measure of the exterior angle with the same vertex is $180 - 144 = 36$.

- Thus, $36 = \dfrac{360}{n}$, so $n = \dfrac{360}{36} = 10$.

The regular polygon has **10** sides.

12. In an isosceles triangle, the base angles are congruent:

$$\text{m}\angle A = \text{m}\angle C = 5x.$$

- Since corresponding angles of congruent triangles are congruent:

$$\text{m}\angle A = \text{m}\angle C$$
$$5x = 2x + 18$$
$$5x - 3x = 18$$
$$\frac{3x}{3} = \frac{18}{3}$$
$$\text{m}\angle A = 5x = 5(6) = 30, \text{ so m}\angle C = 30.$$

- Because the sum of the measures of the angle of a triangle is 180,

$$30 + m\angle B + 30 = 180$$
$$m\angle B = 180 - 60$$
$$= 120$$

- As angles BAG and BAC are supplementary, $m\angle BAG = 180 - 30$ = **150**.

13. In a triangle, the length of each side must be less than the sum of the lengths of the other two sides. Consider each choice in turn until you find a set of numbers that contradicts this fact. In choice (2), $11 \not< 5 + 5$.

The correct choice is **(2)**.

14. Since a hexagon is a polygon with 6 sides, the sum of the measures of its interior angles is $(6 - 2) \times 180 = 720$. It is given that the measures of five of the interior angles of a hexagon are 150, 100, 80, 165, and 150. The sum of the measures of these five interior angles is 645. Hence, the measure of the sixth angle is $720 - 645 = 75$.

The correct choice is **(1)**.

3. PARALLELOGRAMS AND TRAPEZOIDS

3.1 PARALLELOGRAMS

A **parallelogram** is a quadrilateral in which both pairs of opposite sides are parallel. In a parallelogram:

- Opposite sides are congruent.

- Opposite angles are congruent.

- Diagonals bisect each other.

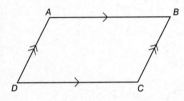   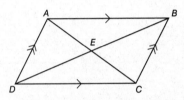

$\overline{AB} \cong \overline{CD}$ and $\overline{AD} \cong \overline{BC}$ $\overline{AE} \cong \overline{EC}$ and $\overline{BE} \cong \overline{ED}$
$\angle A \cong \angle C$ and $\angle B \cong \angle D$

3.2 RECTANGLE

A **rectangle** is a parallelogram with four right angles. The diagonals of a rectangle have the same length.

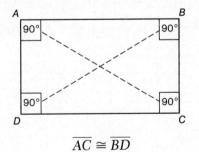

$\overline{AC} \cong \overline{BD}$

3.3 RHOMBUS

A **rhombus** is a parallelogram in which all four sides have the same length. In a rhombus:

- The diagonals intersect at right angles.

- The diagonals bisect the angles at opposite corners of the rhombus.

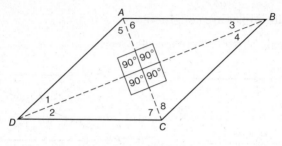

$$\angle 1 \cong \angle 2, \angle 3 \cong \angle 4, \angle 5 \cong \angle 6, \text{ and } \angle 7 \cong \angle 8$$
$$\overline{AB} \cong \overline{BC} \cong \overline{CD} \cong \overline{AD}$$

3.4 SQUARE

A **square** is a parallelogram in which all four sides have the same length and all four angles are right angles. Since a square has all of the properties of a rhombus and a rectangle, in a square:

- The diagonals are congruent and perpendicular.

- Each diagonal bisects the two angles at opposite corners of the square.

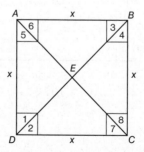

$$\overline{AC} \cong \overline{BD} \text{ and } \overline{AC} \perp \overline{BD}$$
$$\angle 1 \cong \angle 2, \angle 3 \cong \angle 4, \angle 5 \cong \angle 6, \text{ and } \angle 7 \cong \angle 8$$

Property	Rectangle	Rhombus	Square
• All the properties of a parallelogram	Yes	Yes	Yes
• Four right angles	Yes		Yes
• Four congruent sides		Yes	Yes
• Congruent diagonals	Yes		Yes
• Diagonals bisect opposite angles		Yes	Yes
• Diagonals are perpendicular		Yes	Yes

3.5 TRAPEZOIDS

A **trapezoid** is a quadrilateral in which exactly one pair of sides are parallel. The parallel sides are called **bases**, and the nonparallel sides are called **legs**.

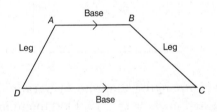

In an isosceles trapezoid:

• Legs are congruent.

• Base angles are congruent.

• Diagonals are congruent.

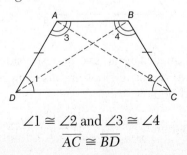

$$\angle 1 \cong \angle 2 \text{ and } \angle 3 \cong \angle 4$$
$$\overline{AC} \cong \overline{BD}$$

Practice Exercises

1. As shown in the accompanying diagram, a rectangular gate has two diagonal supports. If m∠1 = 42, what is m∠2?

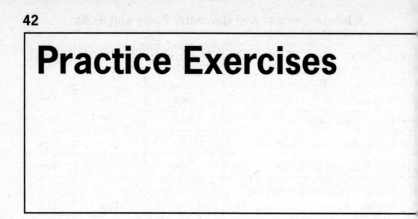

2. In the accompanying diagram of parallelogram *ABCD*, m∠B = 5x and m∠C = 2x + 12. Find the number of degrees in ∠D.

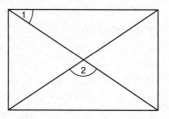

3. In a certain quadrilateral, two opposite sides are parallel, and the other two opposite sides are *not* congruent. This quadrilateral could be a

(1) rhombus   (3) square

(2) parallelogram   (4) trapezoid

4. Which quadrilateral must have diagonals that are congruent and perpendicular?
(1) rhombus   (3) trapezoid
(2) square   (4) parallelogram

5. In the accompanying diagram of parallelogram $ABCD$, diagonals $\overline{AC}$ and $\overline{BD}$ interesect at E, $BE = \frac{2}{3}x$, and $ED = x - 10$.

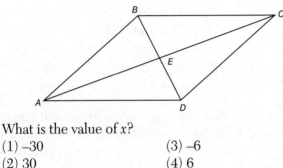

What is the value of x?
(1) –30   (3) –6
(2) 30   (4) 6

Solutions

1.

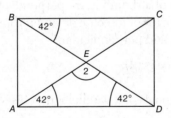

- Since opposite sides of a parallelogram are parallel, m$\angle ADE$ = m$\angle 1$ = 42.

- The diagonals of a rectangle are congruent and bisect each other, so $\triangle AEB$ is isosceles with $\overline{AE} \cong \overline{DE}$. Because base angles of an isosceles triangle are congruent, m$\angle DAE$ = 42.

- Hence,

$$42 + 42 + m\angle 2 = 180$$
$$84 + m\angle 2 = 180$$
$$m\angle 2 = 180 - 84$$
$$= 96$$

2. Since consecutive angles of a parallelogram are supplementary:

$$(2x + 12) + 5x = 180$$
$$7x + 12 = 180$$
$$7x = 180 - 12$$
$$\frac{7x}{7} = \frac{168}{7}$$
$$x = 24$$

3. Since it is given that two opposite sides are not congruent, the quadrilateral cannot be a parallelogram or a special type of parallelogram. A trapezoid has exactly one pair of parallel sides. The other pair of sides may or may not be congruent.

The correct choice is **(4)**.

4. The diagonals of a rhombus are perpendicular. Since a square has all of the properties of a rectangle and a rhombus, its diagonals are both congruent and perpendicular.

The correct choice is **(2)**.

5. Since the diagonals of a parallelogram bisect each other:

$$\frac{2}{3}x = x - 10$$
$$\overset{1}{\cancel{3}}\left(\frac{2}{\cancel{3}}x\right) = 3(x - 10)$$
$$2x = 3x - 30$$
$$30 = 3x - 2x$$
$$x = 30$$

The correct choice is **(2)**.

4. SPECIAL RIGHT TRIANGLE RELATIONSHIPS

4.1 PYTHAGOREAN THEOREM

In a right triangle, the sum of the squares of the lengths of the two legs is equal to the square of the hypotenuse.

PYTHAGOREAN THEOREM: $(\text{side I})^2 + (\text{side II})^2 = (\text{hypotenuse})^2$

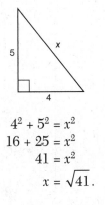

$$4^2 + 5^2 = x^2$$
$$16 + 25 = x^2$$
$$41 = x^2$$
$$x = \sqrt{41}.$$

A set of three positive integers {a,b,c} that satisfies the relationship $a^2 + b^2 = c^2$ is called a **Pythagorean triple**. Common Pythagorean triples include

3-4-5, 5-12-13, and 8-15-17,

where in each case the largest number represents the length of the hypotenuse of the right triangle. Any whole-number multiple of a Pythagorean triple is also a Pythagorean triple.

4.2 MEDIAN TO HYPOTENUSE

In a right triangle, the median drawn to the hypotenuse is half the length of the hypotenuse, as illustrated in the accompanying figure.

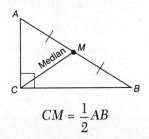

$$CM = \frac{1}{2}AB$$

4.3 30-60 AND 45-45 RIGHT TRIANGLES

30-60 Right Triangle Relationships	45-45 Right Triangle Relationships
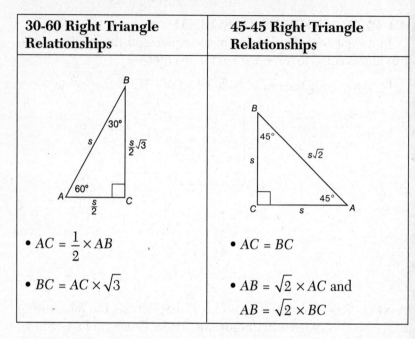	
• $AC = \dfrac{1}{2} \times AB$ • $BC = AC \times \sqrt{3}$	• $AC = BC$ • $AB = \sqrt{2} \times AC$ and $AB = \sqrt{2} \times BC$

4.4 RIGHT TRIANGLE PROPORTIONS

If in a right triangle the altitude to the hypotenuse is drawn, then each of the triangles thus formed is similar to the original triangle and they are similar to each other. Because of the these similar triangles, the following proportions are true:

• $\dfrac{x}{a} = \dfrac{a}{c}$

• $\dfrac{y}{b} = \dfrac{b}{c}$

• $\dfrac{x}{h} = \dfrac{h}{y}$

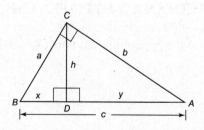

Practice Exercises

1. In right triangle ABC, altitude $\overline{CD}$ is drawn to hypotenuse $\overline{AB}$. If $AD = 2$ and $DB = 6$, than AC is

(1) $4\sqrt{3}$ (3) 3

(2) $2\sqrt{3}$ (4) 4

2. In right triangle ABC, altitude $\overline{CD}$ is drawn to hypotenuse $\overline{AB}$. If $AD = 5$ and $DB = 24$, what is the length of $\overline{CD}$?

(1) 120   (3) $2\sqrt{30}$

(2) $\sqrt{30}$ (4) $4\sqrt{30}$

3. In $\triangle ABC$, $\overline{AB} \perp \overline{BC}$ and $\overline{DE} \perp \overline{CA}$. If $DE = 8$, $CD = 10$, and $CA = 30$, find AB.

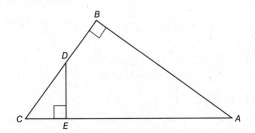

4. Four streets in a town are illustrated in the accompanying diagram. If the distance on Poplar Street from F to P is 12 miles and the distance on Maple Street from E to M is 10 miles, find the distance on Maple Street, in miles, from M to P.

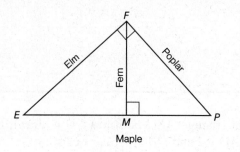

Solutions

1. If an altitude is drawn to the hypotenuse of a right triangle, the length of either leg is the mean proportional between the hypotenuse and the segment of the hypotenuse that is adjacent to that leg. Thus, since $AD = 2$ and $DB = 6$:

$$\frac{AB}{AC} = \frac{AC}{AD}$$

$$\frac{2+6}{AC} = \frac{AC}{2}$$

$$\frac{8}{AC} = \frac{AC}{2}$$

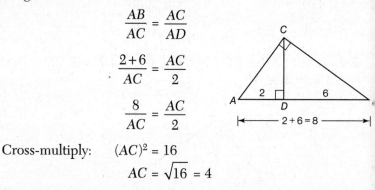

Cross-multiply: $(AC)^2 = 16$

$$AC = \sqrt{16} = 4$$

The correct choice is **(4)**.

2.

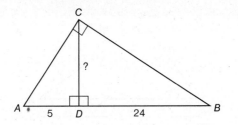

In a right triangle, the length of the altitude drawn to the hypotenuse is the mean proportional between the lengths of the two segments of the hypotenuse formed by the altitude:

$$\frac{AD}{CD} = \frac{CD}{DB}$$

Substitute the given values $AD = 5$ and $DB = 24$ and let $CD = x$:

$$\frac{5}{x} = \frac{x}{24}$$

In a proportion, the product of the mean equals the product of the extremes (cross-multiply):

$$x \cdot x = 5 \cdot 24$$
$$x^2 = 120$$

Take the square root of each side of the equation:

$$x = \pm\sqrt{120}$$

Reject the negative value of x since x represents the length of a side of the triangle:

$$x = \sqrt{120}$$

Factor the radicand so that one of its factors is the highest perfect square factor of 120:

$$x = \sqrt{4 \cdot 30}$$

Distribute the radical sign over each factor of the radicand:

$$x = \sqrt{4} \cdot \sqrt{30}$$

Evaluate $\sqrt{4}$:

$$= 2\sqrt{30}$$

The correct choice is (**3**).

3.

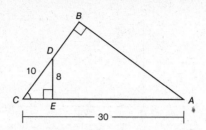

Consider right triangles DEC and ABC. Since $\angle E \cong \angle B$ (right angles are congruent) and $\angle C \cong \angle C$ (reflexive property of congruence), these triangles are similar by the AA $\cong$ AA theorem of similarity.

The lengths of corresponding sides of similar triangles are in proportion:

$$\frac{\text{side in } \triangle ABC}{\text{corresponding side in } \triangle DEC} = \frac{AB}{DE} = \frac{CA}{CD}$$

$$\frac{AB}{8} = \frac{30}{10}$$

In a proportion, the product of the means equals the product of the extremes (cross-multiply):

$$10(AB) = 8(30)$$

$$AB = \frac{240}{10} = 24$$

The length of $\overline{AB}$ is **24**.

4. Represent MP by x:

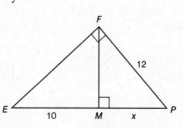

Then:

$$\frac{\text{hypotenuse segment adjacent to } FP}{\text{leg } FP} = \frac{\text{leg } FP}{\text{hypotenuse}}$$

$$\frac{x}{12} = \frac{12}{x+10}$$

$$x(x + 10) = 12 \cdot 12$$

$$x^2 + 10x - 144 = 0$$

$$(x - 8)(x + 18) = 0$$

$$x - 8 = 0 \quad or \quad x + 18 = 0$$
$$x = 8 \qquad\qquad x = -18 \quad \leftarrow \text{reject}$$

The distance from M to P is **8 miles**.

5. LOCUS AND CONCURRENCY THEOREMS

5.1 BASIC LOCI

A **locus** (plural, "loci") is defined as the set of points, and only those points, that satisfy a given set of conditions. To help determine a locus, draw a diagram.

- The locus of points at a fixed distance of d units from a point P is a circle with center at P and radius of length d.

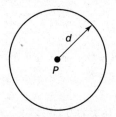

The locus consists of the points on
the circumference of the circle.

- The locus of points at a fixed distance of d units from a given line is a pair of parallel lines with each line d units from the given line.

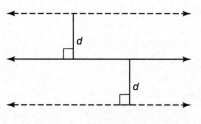

The locus consists of the points on
either of the dotted lines.

• The locus of points equidistant from two fixed points is the perpendicular bisector of the line segment whose endpoints are the fixed points.

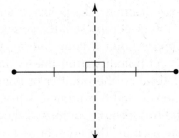

The locus consists of the points on the dotted line.

• The locus of points equidistant from two parallel lines is the parallel line midway between them.

The locus consists of the points on the dotted line.

• The locus of points equidistant from two intersecting lines is the bisector of each pair of vertical angles.

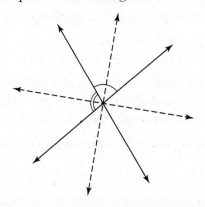

The locus consists of the points on the dotted line.

5.2 COMPOUND LOCI

A locus that involves two or more conditions is called a **compound locus**. To find a compound locus, represent each locus condition on the same diagram. The points, if any, at which all of the locus conditions intersect represent the compound locus.

For example, suppose point P is on line m. What is the total number of points 3 centimeters from line m and 5 centimeters from point P? To determine the number of points 3 centimeters from line m (condition 1) and 5 centimeters from point P (condition 2), represent the two locus conditions as shown in the accompanying diagram. Then count the number of points at which the loci intersect.

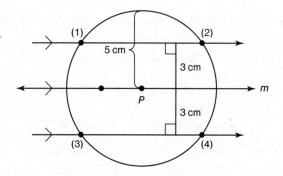

- The locus of points 3 centimeters from line m is a pair of lines, each parallel to line m and each 3 centimeters from line m.

- The locus of points 5 centimeters from point P is a circle with center at P and a radius length of 5 centimeters.

- Since the radius of the circle is greater than 3, the circle intersects each of the parallel lines at two points. Therefore, the two locus conditions intersect at a total of four points.

The total number of points that satisfy the given conditions is **4**.

5.3 MIDPOINT AND CENTROID RELATIONSHIPS

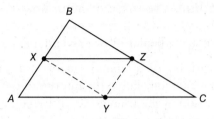

- If X and Z are midpoints, then $\overline{XZ} \parallel \overline{AC}$ and $XZ = \frac{1}{2}AC$.

- If X, Y, and Z are midpoints, then perimeter $\triangle XYZ = \frac{1}{2}$ perimeter of $\triangle ABC$.

5.4 SIDE-SPLITTING THEOREM

If a line intersects two sides of a triangle and is parallel to the third side, then it divides those two sides proportionally. If $\overline{XZ} \parallel \overline{AC}$ in the accompanying figure, then sides $\overline{AB}$ and $\overline{BC}$ are divided proportionally:

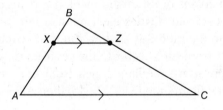

- $\dfrac{BX}{AB} = \dfrac{BZ}{BC}$ or

- $\dfrac{AX}{AB} = \dfrac{CZ}{BC}$ or

- $\dfrac{BX}{AX} = \dfrac{BZ}{CZ}$ or any equivalent proportion.

5.5 CONCURRENCY THEOREMS

Concurrent lines are three or more lines that meet at the same point. A triangle has four types of concurrent lines: altitudes, angle bisectors, perpendicular bisectors, and medians.

- The three altitudes of a triangle meet at a point called the **orthocenter** which may fall in the interior of the triangle (acute triangle), on a side of the triangle (right triangle), or in the exterior of the triangle (obtuse triangle).

- The three angle bisectors of a triangle meet at a point in the interior of the triangle called the **incenter**. Because the sides of the triangle are equidistant from the incenter, the incenter of the triangle represents the center of the circle that can be inscribed in it.

- The perpendicular bisectors of the three sides of a triangle meet at a point called the **circumcenter**. Because the vertices of the triangle are equidistant from the circumcenter, the circumcenter of the triangle represents the center of the circle that can be circumscribed about it.

- The three medians of a triangle meet at a point called the **centroid**. The centroid divides each median into segments whose lengths are in the ratio 2:1, as shown in the accompanying figure. The distance from each vertex to the centroid is two-thirds of the length of the entire median drawn from that vertex. If in the accompanying figure, median $AX = 15$, then $AP = 10$ and $PX = 5$.

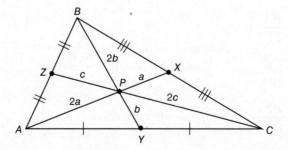

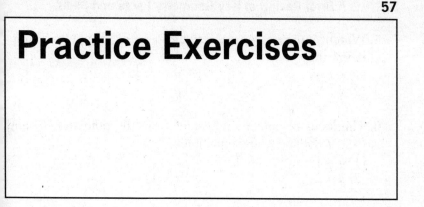

Practice Exercises

1. How many points are equidistant from two intersecting lines ℓ and m and 3 units from the point of intersection of the lines?
   (1) 1   (3) 3
   (2) 2   (4) 4

2. What is the number of inches in the perimeter of a triangle formed by connecting the midpoints of a triangle whose sides measure 5 in., 12 in., and 13 in.?
   (1) 7.5   (3) 30
   (2) 15   (4) 45

3. To locate a point equidistant from the vertices of a triangle, construct
   (1) the perpendicular bisectors of the sides
   (2) the angle bisectors
   (3) the altitudes
   (4) the medians

4. How many points are equidistant from two parallel lines and also equidistant from two points on one of the lines?
   (1) 1   (3) 3
   (2) 2   (4) 4

5. Which equation represents the locus of all points 5 units below the x-axis?
(1) $x = -5$ (3) $y = -5$
(2) $x = 5$ (4) $y = 5$

6. The locus of points equidistant from the points $(4,-5)$ and $(4,7)$ is the line whose equation is
(1) $y = 1$ (3) $x = 1$
(2) $y = 2$ (4) $x = 4$

7. How many points are 3 units from the origin and 2 units from the y-axis?

8. In the accompanying diagram of $\triangle PRT$, $\overline{KG} \parallel \overline{PR}$. If $TP = 20$, $KP = 4$, and $GR = 7$, what is the length of TG?

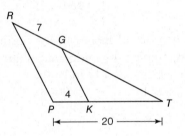

Solutions
  1.

- Locus condition 1: The locus of a point equidistant from two intersecting lines, ℓ and m, is the pair of lines that bisect the vertical pairs of angles formed by the two lines, as indicated by the broken lines in the accompanying figure.

- Locus condition 2: The locus of points 3 units from the point of intersection of the two given lines is a circle whose center is the point of intersection of the lines and whose radius is 3 units.

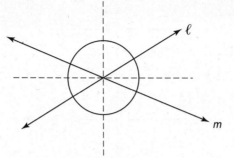

- Since the circle intersects the broken lines in a total of 4 points, there are four points that satisfy both locus conditions.

  The correct choice is **(4)**.

  2. Since the line segment joining the midpoints of two sides of a triangle is one-half the length of the third side, the lengths of the triangle formed by connecting the midpoints of a triangle whose sides measure 5 in., 12 in., and 13 in. are 2.5 in., 6 in., and 6.5 in. The perimeter of the triangle thus formed is 2.5 in. + 6 in. + 6.5 in. = 15 in.

  The correct choice is **(2)**.

  3. The perpendicular bisectors of the three sides of a triangle meet at a point, called the circumcenter, that is equidistant from the vertices of the triangle.

  The correct choice is **(1)**.

  4.
- Locus condition 1: The locus of points equidistant from two parallel lines is a parallel line midway between the two given lines, as represented by the broken line in the accompanying figure.

- Locus condition 2: If A and B are two different points on one of the two given lines, then the locus of points equidistant from these points is the perpendicular bisector of $\overline{AB}$.

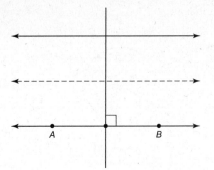

- Since the perpendicular bisector of $\overline{AB}$ intersects the broken line in one point, there is exactly one point that satisfies both locus conditions.

The correct choice is (**1**).

5. The locus of points 5 units below the x-axis is a line parallel to the x-axis which intersects the y-axis at -5. An equation of this line is $y = -5$.

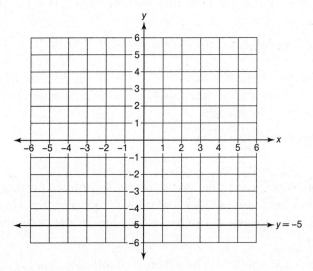

The correct choice is (**3**).

6. The locus of points equidistant from the two given points is the perpendicular bisector of the segment determined by the two points. The midpoint of (4,–5) and (4,7) is (4,1), as shown in the accompanying figure.

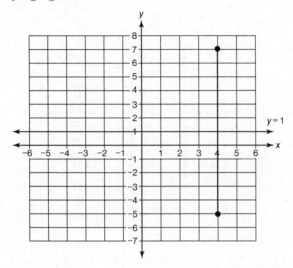

Hence, the perpendicular bisector of the vertical line determined by the two given points is the horizontal through (4,1), an equation of which is $y = 1$.

The correct choice is **(1)**.

7.

- Locus condition 1: All points 3 units from the origin. The desired locus is a circle with the origin as its center and a radius of 3 units. Note in the accompanying diagram that the circle intersects each coordinate axis at 3 and –3.

- Locus condition 2: All points 2 units from the y-axis. The desired locus is a pair of parallel lines; one line is 2 units to the right of the y-axis ($x = 2$), and the other line is 2 units to the left of the y-axis ($x = -2$). See the accompanying diagram.

- Since the loci intersect at points A, B, C, and D, there are four points that satisfy both conditions.

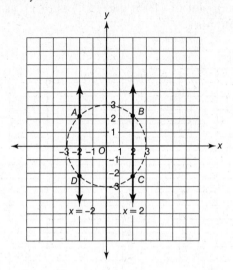

8. Since $\overline{KG} \parallel \overline{PR}$, $\overline{KG}$ divides the sides that it intersects proportionately. Thus:

$$\frac{TG}{GR} = \frac{TK}{KP}$$

$$\frac{TG}{7} = \frac{20-4}{4}$$

$$\frac{TG}{7} = \frac{16}{4}$$

$$\frac{TG}{7} = \frac{4}{1}$$

$$TG = 4 \times 7 = 28$$

The length of $\overline{TG}$ is **28**.

6. CIRCLES AND ANGLE MEASUREMENT

6.1 ARC LENGTH AND AREA OF A SECTOR

A central angle of n degrees intercepts an arc L and a sector such that:

• Arc length $L = \dfrac{n}{360} \times \overset{\text{circumference}}{(2\pi r)}$

• Area sector $AOB = \dfrac{n}{360} \times \overset{\text{area of circle}}{(\pi r^2)}$

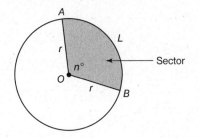

6.2 ARC AND CHORD RELATIONSHIPS

• In the same or congruent circles, congruent chords are equidistant from the center, and chords that are the same distance from the center are congruent:

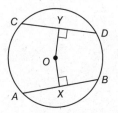

$$\overline{AB} \cong \overline{CD} \Rightarrow OX = OY \text{ and } OX = OY \Rightarrow \overline{AB} \cong \overline{CD}$$

- A diameter perpendicular to a chord bisects the chord and its arcs:

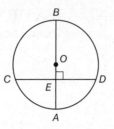

$$\overline{AB} \perp \overline{CD} \Rightarrow \overline{CE} \cong \overline{DE}, \; \overset{\frown}{AC} \cong \overset{\frown}{AD}, \text{ and } \overset{\frown}{BC} \cong \overset{\frown}{BD}$$

- In the same or congruent circles, congruent chords intercept congruent arcs, and congruent arcs have congruent chords:

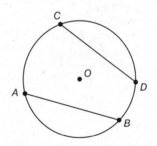

$$\overline{AB} \cong \overline{CD} \Rightarrow \overset{\frown}{AB} \cong \overset{\frown}{CD} \text{ and } \overset{\frown}{AB} \cong \overset{\frown}{CD} \Rightarrow \overline{AB} \cong \overline{CD}$$

- In a circle, parallel chords cut off congruent arcs:

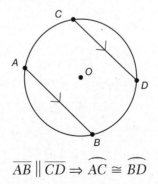

$$\overline{AB} \parallel \overline{CD} \Rightarrow \overset{\frown}{AC} \cong \overset{\frown}{BD}$$

• If two chords intersect in the interior of a circle, then the product of the segment lengths of one chord is equal to the product of the segment lengths of the other chord:

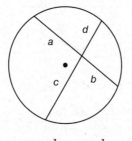

$$a \times b = c \times d$$

6.3 TANGENT AND SECANT RELATIONSHIPS

A **tangent** is a line that intersects a circle at exactly one point called the point of contact or point of tangency. A **secant** is a line that intersects a circle in two distinct points.

• A radius drawn to the point of contact of a tangent is perpendicular to the tangent at that point:

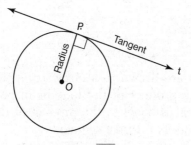

Radius $\overline{OP} \perp t$

• If two tangents are drawn to a circle from the same exterior point, then the tangent segments are congruent. The line segment whose endpoints are the center of the circle and the exterior point bisects the angle formed by the two tangents:

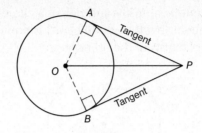

$$\overline{PA} \cong \overline{PB} \text{ and } \overline{OP} \text{ bisects } \angle APB$$

- If a tangent and a secant are drawn to a circle from the same exterior point, then the length of the tangent segment is the mean proportional between the lengths of the secant segment and its external segment:

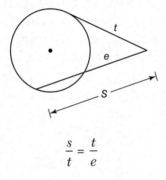

$$\frac{s}{t} = \frac{t}{e}$$

- If two secants are drawn to a circle from the same exterior point, then the product of the lengths of one secant and its external segment is equal to the product of the lengths of the other secant and its external segment:

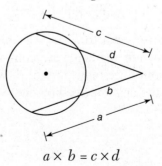

$$a \times b = c \times d$$

5.4 TANGENT CIRCLES

Tangent circles are circles in the same plane that are tangent to the same line at the same point. A line that is tangent to two circles is called a **common tangent**.

- *Internally* tangent circles are tangent circles that lie on the same side of their common tangent:

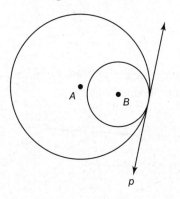

- *Externally* tangent circles are tangent circles that lie on opposite sides of their common tangent:

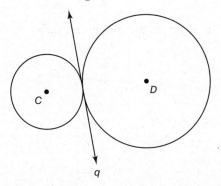

Common tangent lines may be classified as either *internal* or *external*. A common *internal* tangent intersects the line joining the centers of the two circles at a point between the two circles. A common *external* tangent does *not* intersect the line through the centers of the two circles.

- Internally tangent circles have one common external tangent:

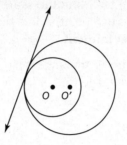

- Externally tangent circles have two common external tangents and one common internal tangent:

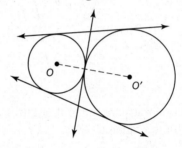

- Circles that intersect at two points have two common external tangents:

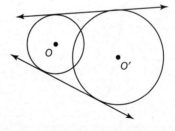

• Nonintersecting circles, one outside the other, have two common external tangents and two common internal tangents:

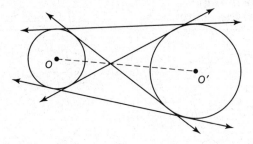

6.5 ANGLE MEASUREMENT RELATIONSHIPS

The formula used to determine the measure of an angle from its intercepted arc(s) depends on the location of the vertex of the angle.

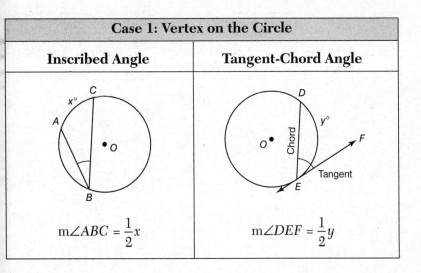

Case 1: Vertex on the Circle	
Inscribed Angle	**Tangent-Chord Angle**
$m\angle ABC = \dfrac{1}{2}x$	$m\angle DEF = \dfrac{1}{2}y$

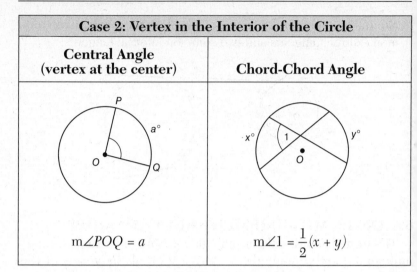

Case 2: Vertex in the Interior of the Circle	
Central Angle **(vertex at the center)**	**Chord-Chord Angle**
$\mathrm{m}\angle POQ = a$	$\mathrm{m}\angle 1 = \dfrac{1}{2}(x + y)$

Case 3: Vertex in the Exterior of the Circle		
Secant-Tangent **Angle**	**Secant-Secant** **Angle**	**Tangent-Tangent** **Angle**
$\mathrm{m}\angle 1 = \dfrac{1}{2}(x - y)$	$\mathrm{m}\angle 1 = \dfrac{1}{2}(x - y)$	$\mathrm{m}\angle 1 = \dfrac{1}{2}(x - y)$

Practice Exercises

1. In the accompanying diagram, $\overrightarrow{BD}$ is tangent to circle O at B, $\overline{BC}$ is a chord, and $\overline{BOA}$ is a diameter. If $m\widehat{AC} : m\widehat{CB} = 1:4$, find m∠DBC.

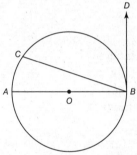

2. In the accompanying diagram of circle O, $\widehat{AB} \cong \widehat{CD}$.

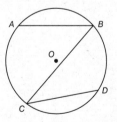

Which statement is true?
(1) $\overline{AB} \cong \overline{CD}$ (3) $\overline{AB} \parallel \overline{CD}$

(2) $\widehat{AC} \cong \widehat{BD}$ (4) $\angle ABC \cong \angle BCD$

3. In the accompanying diagram of circle O, $\overline{AB}$ and $\overline{BC}$ are chords and m∠AOC = 96. What is m∠ABC?

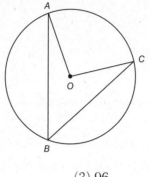

(1) 32
(2) 48

(3) 96
(4) 192

4. Intersecting circles have a maximum of
   (1) one common tangent
   (2) two common tangents
   (3) three common tangents
   (4) four common tangents

5. In the accompanying diagram, $\overline{AB}$ is tangent to circle O at B. If AC = 16 and CD = 9, what is the length of $\overline{AB}$?

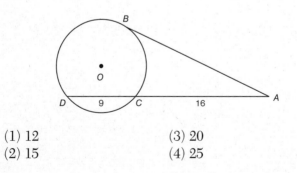

(1) 12
(2) 15

(3) 20
(4) 25

6. In the diagram below, chords $\overline{AB}$ and $\overline{CD}$ intersect at E. If m∠AEC = 4x, m$\overset{\frown}{AC}$ = 120, and m$\overset{\frown}{DB}$ = 2x, what is the value of x?

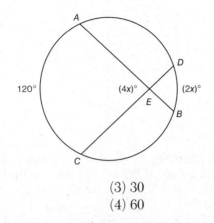

(1) 12   (3) 30
(2) 20   (4) 60

7. As shown in the accompanying diagram, a dial in the shape of a semicircle has a radius of 4 centimeters. Find the measure of θ, correct to the *nearest degree*, when the pointer rotates to form an arc whose length is 1.38 centimeters.

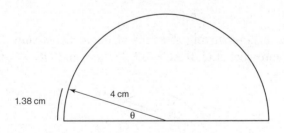

8. In the accompanying diagram of circle O, chords $\overline{AB}$ and $\overline{CD}$ intersect at point E. If $AE = 2$, $CD = 9$, and $CE = 4$, find BE.

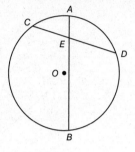

9. In the accompanying diagram, $\overrightarrow{PC}$ is tangent to circle O at C and $\overleftrightarrow{PAB}$ is a secant. If $PC = 8$ and $PA = 4$, find AB.

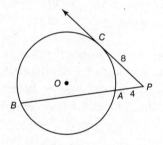

10. In the accompanying diagram of circle O, secants $\overline{CBA}$ and $\overline{CED}$ intersect at C. If $AC = 12$, $BC = 3$, and $DC = 9$, find EC.

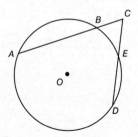

11. In the accompanying diagram, $\overline{AFB}$, $\overline{AEC}$, and $\overline{BGC}$ are tangent to circle O at F, E, and G, respectively. If $AB = 32$, $AE = 20$, and $EC = 24$, find BC.

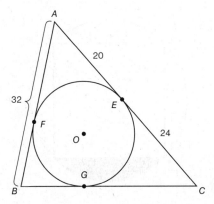

12. In the accompanying diagram of circle O, $\overline{AE}$ and $\overline{FD}$ are chords, $\overline{AOBG}$ is a diameter and is extended to C, $\overline{CDE}$ is a secant, $\overline{AE} \parallel \overline{FD}$, and $\text{m}\widehat{AE} : \text{m}\widehat{ED} : \text{m}\widehat{DG} = 5:3:1$.

Find:

a $\text{m}\widehat{DG}$

b $\text{m}\angle AEF$

c $\text{m}\angle DBG$

d $\text{m}\angle DCA$

e $\text{m}\angle CDF$

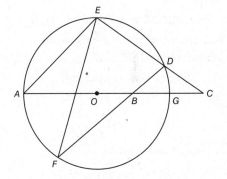

13. A regular hexagon is inscribed in a circle whose area is 144π.
   a. What is the length of the minor arc intercepted by a side of the hexagon?
   b. What is the area of the hexagon?

14. Square $ABCD$ is inscribed in circle O, as shown in the accompanying figure. If the radius is 8 inches, find the area of the shaded region in terms of π.

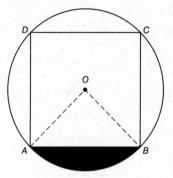

Solutions

1. Let $m\overarc{AC} = x$.

Since $m\overarc{AC} : m\overarc{CB} = 1:4$,

$m\overarc{CB} = 4x$.

Since $\overline{BOA}$ is a diameter, $\overarc{ACB}$ is a semicircle; the measure of a semicircle is 180:

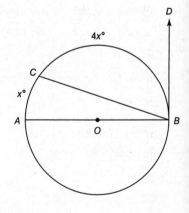

$$m\overarc{ACB} = 180$$
$$x + 4x = 180$$
$$5x = 180$$
$$x = 36$$
$$m\overarc{CB} = 4x = 4(36) = 144$$
$$m\angle DBC = \frac{1}{2}(144) = 72$$

2. Congruent arcs cut off congruent chords. Since it is given that $\overset{\frown}{AB} \cong \overset{\frown}{CD}$, then $\overline{AB} \cong \overline{CD}$.

The correct choice is (**1**).

3. A central angle is measured by its intercepted arc. Since it is given that the measure of central AOC is 96, $\overset{\frown}{AC} = 96$.

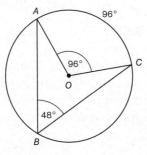

The measure of an inscribed angle is one-half the measure of its intercepted arc. Hence, $m\angle ABC = \dfrac{1}{2}(96) = 48$.

The correct choice is (**2**).

4. Intersecting circles may intersect at one or two points:

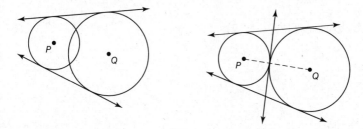

Thus, two intersecting circles may have a maximum of three common tangents.

The correct choice is (**3**).

5. If a tangent and a secant are drawn to a circle from the same point, then the tangent is the mean proportional between the external segment of the secant and the whole secant:

$$\frac{16}{AB} = \frac{AB}{9+16}$$

$$AB \times AB = 16 \times 25$$

$$(AB)^2 = 400$$

$$AB = \sqrt{400} = 20$$

The correct choice is **(3)**.

6. Angle *AEC* is an *angle formed by two chords intersecting within the circle*; the measure of an angle formed by two chords intersecting within a circle is equal to one-half the sum of the measures of the two intercepted arcs:

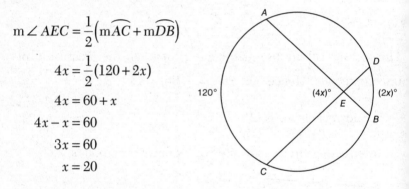

$$m \angle AEC = \frac{1}{2}\left(m\widehat{AC} + m\widehat{DB}\right)$$

$$4x = \frac{1}{2}(120 + 2x)$$

$$4x = 60 + x$$

$$4x - x = 60$$

$$3x = 60$$

$$x = 20$$

The correct choice is **(2)**.

7. It is given that the radius of the circle is 4 cm and the length of the intercepted arc is 1.38 cm. Thus:

$$\text{Arc length } L = \frac{n}{360} \times \overbrace{(2\pi r)}^{\text{circumference}}$$

$$1.38 = \frac{\theta}{360} \times 2\pi(4)$$

$$1.38 = \frac{\theta\pi}{45}$$

$$\frac{138 \times 45}{\pi} = \theta$$

$$\theta = 19.76704393$$

To the *nearest degree*, $\theta = 20$.

8. Let $BE = x$. If two chords intersect within a circle, the product of the lengths of the segments of one chord equals the product of the lengths of the segments of the other chord:

$$DE = CD - CE = 9 - 4 = 5.$$

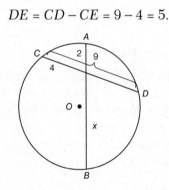

Therefore:

$$AE \times BE = CE \times ED$$
$$2x = 4(5)$$
$$2x = 20$$
$$x = 10$$

$BE = \mathbf{10}$.

9. Let $AB = x$. If a tangent and a secant are drawn to a circle from an outside point, the length of the tangent is the mean proportional between the length of the whole secant and the length of its external segment:

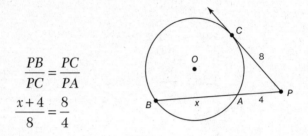

$$\frac{PB}{PC} = \frac{PC}{PA}$$

$$\frac{x+4}{8} = \frac{8}{4}$$

In a proportion, the product of the means equals the product of the extremes (cross-multiply):

$$4(x + 4) = 8(8)$$
$$4x + 16 = 64$$
$$4x = 64 - 16$$
$$4x = 48$$
$$x = 12$$

$AB = \mathbf{12}$.

10. It is given that, in the accompanying diagram of circle O, secants $\overline{CBA}$ and $\overline{CED}$ intersect at C, $AC = 12$, $BC = 3$, and $DC = 9$.

If two secants are drawn to a circle from the same exterior point, the product of the length of one secant and the length of its external segment is equal to the product of the length of the other secant and the length of its external segment.

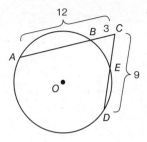

Thus:

$$AC \times BC = DC \times EC$$

Since $AC = 12$, $BC = 3$, and $DC = 9$:

$$12 \times 3 = 9 \times EC$$
$$36 = 9(EC)$$
$$\frac{36}{4} = EC$$
$$4 = EC$$

The length of $\overline{EC}$ is **4**.

11. Tangent segments drawn to a circle from the same point have the same length. Thus, in the accompanying diagram:

$AF = AE = 20$
$BF = 32 - AF = 32 - 20 = 12$
$BG = BF = 12$
$CG = CE = 24$
$BC = BG + CG = 12 + 24 = 36$

$BC = \mathbf{36}$.

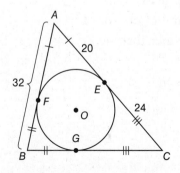

12.

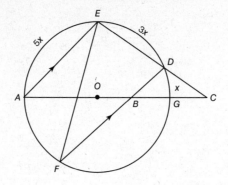

a. Given:

$$m\overset{\frown}{AE} : m\overset{\frown}{ED} : m\overset{\frown}{DG} = 5:3:1$$

Let:

$$m\overset{\frown}{DG} = x$$

Then:

$$m\overset{\frown}{ED} = 3x$$

and

$$m\overset{\frown}{AE} = 5x$$

Since $\overline{AOBG}$ is a diameter, $\overset{\frown}{AEDG}$ is a semicircle, and $m\overset{\frown}{AEDG} = 180$:

$$m\overset{\frown}{AE} + m\overset{\frown}{ED} + m\overset{\frown}{DG} = \overset{\frown}{AEDG}$$
$$5x + 3x + x = 180$$
$$9x = 180$$
$$x = 20$$

$m\overset{\frown}{DG} = \mathbf{20}$.

b. From part **a**, $m\overset{\frown}{ED} = 3x$. $\overline{AE} \parallel \overline{FD}$, and parallel chords intercept on circle arcs that are equal in measure:

$$m\overset{\frown}{AF} = m\overset{\frown}{ED} = 3x$$

From part **a**, $x = 20$:

$$m\overset{\frown}{AF} = 3(20) = 60$$

Angle *AEF* is an *inscribed angle*; the measure of an inscribed angle is equal to one-half the measure of its intercepted arc:

$$\text{m}\angle AEF = \frac{1}{2}\text{m}\,\widehat{AF}$$

$$= \frac{1}{2}(60) = 30$$

m∠*AEF* = **30**.

c. Angle *DBG* is an angle formed by two chords intersecting within the circle; the measure of an angle formed by two chords intersecting within a circle is equal to one-half the sum of the measures of the intercepted arcs:

$$\text{m}\angle DBG = \frac{1}{2}(\text{m}\,\widehat{AF} + \text{m}\,\widehat{DG})$$

From part **a**, m$\widehat{DG}$ = 20; from part **b**, m$\widehat{AF}$ = 60:

$$\text{m}\angle DBG = \frac{1}{2}(60 + 20)$$

$$= \frac{1}{2}(80) = 40$$

m∠*DBG* = **40**.

d. Angle *DCA* is an angle formed by two secants intersecting outside the circle; the measure of an angle formed by two secants intersecting outside a circle is equal to one-half the difference of the measures of the intercepted arcs:

$$\text{m}\angle DCA = \frac{1}{2}(\text{m}\,\widehat{AE} - \text{m}\,\widehat{DC})$$

From part **a**, m$\widehat{DG}$ = 20; also
m$\widehat{AE}$ = 5x and x = 20; therefore,
m$\widehat{AE}$ = 5(20) = 100:

$$m\angle DCA = \frac{1}{2}(100 - 20)$$

$$= \frac{1}{2}(80)$$

$$= 40$$

m$\angle DCA$ = **40**.

e. Angle *EDF* is an inscribed angle; the measure of an inscribed angle is equal to one-half the measure of its intercepted arc:

$$m\angle EDF = \frac{1}{2}m\,\widehat{FAE}$$

$$m\,\widehat{FAE} = m\,\widehat{AF} + m\,\widehat{AE}$$

From part **b**, m$\widehat{AF}$ = 60; from part **d**, m$\widehat{AE}$ = 100:

$$m\,\widehat{FAE} = 60 + 100 = 160$$

$$m\angle EDF = \frac{1}{2}(160) = 80$$

Angle *CDF* is the supplement of $\angle EDF$; the sum of the measures of two supplementary angles is 180:

$$m\angle CDF + m\angle EDF = 180$$

$$m\angle CDF + 80 = 180$$

$$m\angle CDF = 180 - 80 = 100$$

m$\angle CDF$ = **100**.

13a.

- If the area of the circle = 144π, then $r^2 = 144$, so $r = \sqrt{144} = 12$. Hence, the circumference of the circle is $2 \times \pi \times 12 = 24\pi$.

- The six congruent sides of a regular hexagon divide the circumference of a circle into six congruent arcs.

- Hence, the length of each minor intercepted arc is $\frac{1}{6} \times 24\pi = 4\pi$.

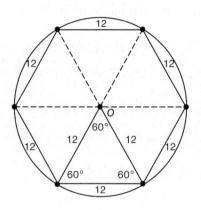

b.

- Drawing radii to the vertices of an inscribed regular hexagon, as shown in the accompanying figure, forms six congruent equilateral triangles in which each central angle measures $\frac{360°}{6} = 60$.

- Hence, the length of each side of the regular hexagon is equal to 12, the radius of the circle.

- To find the area of an equilateral triangle with side s, use the formula $\frac{s^2}{4} \times \sqrt{3}$:

$$\text{area} = \frac{12^2}{4} \times \sqrt{3} = \frac{144}{4}\sqrt{3} = 36\sqrt{3}.$$

- Since a regular hexagon is composed of six congruent triangles:

  area of regular hexagon = $6 \times 36\sqrt{3} = 216\sqrt{3}$.

14. Draw radii $\overline{OA}$ and $\overline{OB}$. To find the area of the shaded region, subtract the area of $\angle AOB$ from the area of sector AOB.

- Since $\overset{\frown}{AB}$ is one-fourth of the circumference of the circle, $\text{m}\overset{\frown}{AB} = \dfrac{1}{4} \times 360 = 90$. Therefore, $\angle AOB$ is a right angle.

- The area of circle O is $\pi \times 8^2 = 64\pi$ in.^{2} Since sector AOB is one-fourth of the circle, the area of sector AOB is $\dfrac{1}{4} \times 64\pi = 16\pi$ in.^{2}

- The area of isosceles right triangle AOB is

  $$\frac{1}{2} \times OA \times OB = \frac{1}{2} \times 8 \times 8 = 32 \text{ in.}^2$$

- The area of the shaded region is $16\pi - 32$ in.^{2}

7. COORDINATE RELATIONSHIPS

7.1 MIDPOINT, SLOPE, AND DISTANCE FORMULAS

Given points $A(x_A, y_A)$ and $B(x_B, y_B)$:

- Midpoint of $\overline{AB} = \left(\dfrac{x_A + x_B}{2}, \dfrac{y_A + y_B}{2} \right)$

- Slope of $\overline{AB} = m = \dfrac{\Delta y}{\Delta x} = \dfrac{y_B - y_A}{x_B - x_A}$

- Length of $\overline{AB} = \sqrt{(x_B - x_A)^2 + (y_B - y_A)^2}$

7.2 SOME FACTS ABOUT SLOPE

- The slope of a horizontal line is 0, and the slope of a vertical line is not defined.

- If two nonvertical lines are parallel, they have the same slope. Conversely, if two lines have the same slope, they are parallel.

- If two nonvertical lines are perpendicular, the product of their slopes is –1 (negative reciprocals). Conversely, if the product of the slopes of two lines is –1 (negative reciprocals), they are perpendicular.

7.3 EQUATION OF A LINE

An equation of a line represents the general relationship between the x-coordinates and the corresponding y-coordinates of all points that lie on the line.

- If m is the slope of a line that crosses the y-axis at $y = b$, then an equation of this line is

$$y = mx + b.$$

- If m is the slope of a line that contains the point (x_A, y_A), then an equation of this line is

$$y - y_A = m(x - x_A).$$

7.4 EQUATION OF A CIRCLE

The center-radius form of an equation of a circle represents the general relationship between the x- and the corresponding y-coordinates of all points that lie on the circumference of the circle of a given radius and center. If the center of a circle is (h,k) and its radius is r, then its equation is

$$(x - h)^2 + (y - k)^2 = r^2.$$

- If the center of a circle with a radius of 6 is at $(2,-3)$, then its equation is

$$(x - 2)^2 + (y - (-3))^2 = 6^2,$$

or, equivalently,

$$(x - 2)^2 + (y + 3)^2 = 36.$$

- If an equation of a circle is $(x + 1)^2 + (y - 4)^2 = 81$, then by rewriting it in the center-radius form $(x - h)^2 + (y - k)^2 = r^2$, you can determine the coordinates of its center and the radius length:

$$\left(x - \underbrace{(-1)}_{h}\right)^2 + \left(y - \underbrace{(-4)}_{k}\right)^2 = \underbrace{9}_{r}{}^2.$$

Thus, the center is at $(-1,4)$ and the radius is 9.

Practice Exercises

1. Segment RS is parallel to segment TU. If the slope of $\overline{RS} = \dfrac{5}{8}$ and the slope of $\overline{TU} = \dfrac{x}{24}$, the value of x is

   (1) 20   (3) 10
   (2) 15   (4) 5

2. The diagonals of parallelogram $ABCD$ intersect at point E. If the coordinates of A and C are $(4,-2)$ and $(-2,5)$, respectively, what are the coordinates of point E?

   (1) $\left(1,\dfrac{3}{2}\right)$ (3) $\left(1,\dfrac{7}{2}\right)$

   (2) $\left(\dfrac{3}{2},3\right)$ (4) $\left(2,\dfrac{3}{2}\right)$

3. Which equation represents the locus of a point 4 units from the point $(-1,3)$?

   (1) $(x-1)^2 + (y+3)^2 = 4$
   (2) $(x+1)^2 + (y-3)^2 = 4$
   (3) $(x-1)^2 + (y+3)^2 = 16$
   (4) $(x+1)^2 + (y-3)^2 = 16$

4. The vertices of triangle ABC are $A(1,2)$, $B(9,6)$, and $C(5,2)$. What is an equation of the median to $\overline{AC}$?

(1) $y = \dfrac{3}{2}x$ (3) $y = \dfrac{2}{3}x$

(2) $y = \dfrac{3}{2}x + 1$ (4) $y = \dfrac{2}{3}x + 1$

5. The vertices of triangle RST are $R(1,3)$, $S(7,1)$, and $T(5,4)$. What is an equation of the altitude from T to $\overline{RS}$?

(1) $y = -\dfrac{1}{3}x$ (3) $y = 3x - 11$

(2) $y = -\dfrac{1}{3}x - 11$ (4) $y = 3x$

Solutions

1. Since parallel lines have the same slope,

$$\text{slope of } \overline{RS} = \text{slope of } \overline{TU}$$

$$\frac{5}{8} = \frac{x}{24}$$

$$8x = 5(24)$$

$$\frac{8x}{8} = \frac{120}{8}$$

$$x = 15$$

The correct choice is **(2)**.

2. The diagonals of a parallelogram bisect each other, so that their point of intersection is the midpoint of each diagonal. To determine the coordinates of point E, find the coordinates of the midpoint of diagonal $\overline{AC}$. If $(\bar{x}, \bar{y})$ are the coordinates of point E, then:

$$\bar{x} = \frac{4 + (-2)}{2} = \frac{2}{2} = 1 \quad \text{and} \quad \bar{y} = \frac{-2 + 5}{2} = \frac{3}{2}.$$

The coordinates of point E are $\left(1, \frac{3}{2}\right)$.

The correct choice is (**1**).

3. The locus of all points 4 units from the point $(-1,3)$ is a circle whose center is at $(-1,3)$ and whose radius is 4. To find an equation of this locus, use the center-radius form of an equation of a circle where $r = 4$ and $(h,k) = (-1,3)$:

$$(x - h)^2 + (y - k)^2 = r^2$$
$$(x - (-1))^2 + (y - 3)^2 = 4^2$$
$$(x + 1)^2 + (y - 3)^2 = 16$$

The correct choice is (**4**).

4. A median of a triangle is a segment drawn from one of the vertices of the triangle to the midpoint of the opposite side.

- The median to $\overline{AC}$, intersects $\overline{AC}$ at its midpoint, which is at

$$M\left(\frac{5+1}{2}, \frac{2+2}{2}\right) = M(3,2).$$

- The slope m of median $\overline{BM}$ is

$$m = \frac{\Delta y}{\Delta x} = \frac{6-2}{9-3} = \frac{4}{6} = \frac{2}{3}.$$

- To find an equation of the median, use the point-slope form of the equation of a line where $m = \frac{2}{3}$ and $(x_a, y_a) = (3,2)$:

$$y - y_A = m\,(x - x_A)$$
$$y - 2 = \frac{2}{3}(x - 3)$$

- Rewrite the equation in slope-intercept form:

$$y - 2 = \frac{2}{3}x + \frac{2}{3}(-3)$$

$$y - 2 = \frac{2}{3}x - 2$$

$$y = \frac{2}{3}x$$

The correct choice is (3).

5. An altitude of a triangle is a segment drawn from one of the vertices of the triangle and perpendicular to the opposite side.

- The slope m of $\overline{RS}$ is

$$m = \frac{\Delta y}{\Delta x} = \frac{1-3}{7-1} = \frac{-2}{6} = -\frac{1}{3}.$$

- Since an altitude intersects the side to which it is drawn at right angles, the slopes of $\overline{RS}$ and the altitude to $\overline{RS}$ must be negative reciprocals. Hence, the slope of the altitude to $\overline{RS}$ is 3.

- To find an equation of the altitude, use the point-slope form of the equation of a line where $m = 3$ and $(x_a, y_a) = (5,4)$:

$$y - y_A = m(x - x_A)$$
$$y - 4 = 3(x - 5)$$

- Rewrite the equation in slope-intercept form:

$$y - 4 = 3x - 15$$
$$y = 3x - 11$$

The correct choice is (3).

8. VOLUME AND SURFACE AREA

Let h = height, ℓ = slant height, p = perimeter, r = radius, and B = area of a base.

Solid Figure*	Volume (V)	Area†
• Prism	$V = B \times h$	L.A. = $h\,p$
• Pyramid	$V = \dfrac{1}{3}B \times h$	L.A. = $\dfrac{1}{2}p\ell$
• Cylinder	$V = \pi r^2 h$	L.A. = $2\pi rh$
• Cone	$V = \dfrac{1}{3}\pi r^2 h$	L.A. = $\pi r\ell$
• Sphere	$V = \dfrac{4}{3}\pi r^3$	S.A. = $4\pi r^2$

*Volume formulas hold for both right and oblique solids (prisms, cylinders, and cones) and for both regular and nonregular pyramids.
†L.A. = lateral area; S.A. = surface area.

Practice Exercises

1. A prism has a right triangular base as shown in the accompanying figure. Find the volume of the prism.

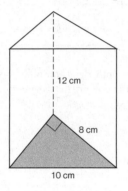

12 cm

8 cm

10 cm

2. In the accompanying diagram of a pyramid with square base *ABCD*, *BC* = 18 cm and slant height *JL* = 15 cm. Find the lateral area and the volume of the pyramid.

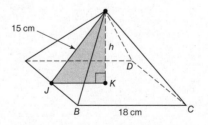

15 cm

h

D

K

J

B   18 cm   *C*

3. In the accompanying figure showing a right cone, the vertex angle measures 60° and the slant height is 18 cm. Find, to the *nearest tenth* of a cubic centimeter, the volume of the cone.

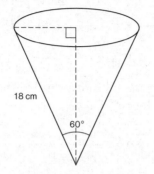

Solutions

1. The volume of a prism is the product of the area of its base and its height. The lengths of the sides of the right triangular base of the given figure form a 6-8-10 Pythagorean triple.

- The area of the right triangle base of the prism is $\frac{1}{2} \times 6$ cm $\times 8$ cm $= 24$ cm^2.

- Since the height of the prism is given as 12 cm,

$$\text{volume} = 24 \text{ cm}^2 \times 12 \text{ cm} = 288 \text{ cm}^3.$$

2. As the base of the given pyramid is a square with center K,

$$JK = \frac{1}{2} \times 18 \text{ cm} = 9 \text{ cm}.$$

The lengths of the sides of right triangle JKL form a 9-12-15 Pythagorean triple in which $h = 12$ cm and the slant height is 15 cm.

- The lateral area (L.A.) of a pyramid is one-half of the product of its slant height and the perimeter of its base. Thus:

$$\text{L.A.} = \frac{1}{2} \times 15 \text{ cm} \times \overbrace{(4 \times 18 \text{ cm})}^{\text{perimeter of square base}}$$

$$= 540 \text{ cm}^2$$

- The volume of a pyramid is one-third of the product of the area of its base and its height. Thus:

$$\text{volume} = \frac{1}{3} \times \overbrace{(18 \text{ cm} \times 18 \text{ cm})}^{\text{area of square base}} \times 15 \text{ cm}$$

$$= 1620 \text{ cm}^2$$

3.

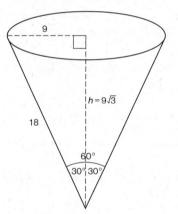

In a 30-60 right triangle, the side opposite the 30° angle is one-half of the hypotenuse, and the side opposite the 60° angle is $\sqrt{3}$ times the length of the shorter leg. As shown in the accompanying figure, the radius of the base of the right cone is 9 cm, and the height of the cone is $9\sqrt{3}$ cm. The volume V of a right circular cone is given by the formula $V = \dfrac{1}{3}\pi r^2 h$, where r is the radius of the circular base and h is the height of the cone. Since $r = 9$ cm and $h = 9\sqrt{3}$ cm:

$$V = \frac{1}{3}\pi(9 \text{ cm})^2(9\sqrt{3} \text{ cm})$$

$$= 243\pi\sqrt{3} \text{ cm}^3$$

$$= 243 \times \pi \times \sqrt{3} \text{ cm}^3$$

Use your calculator: $= 1322.259737 \text{ cm}^3$

The volume of the cone, correct to the *nearest tenth of a cubic centimeter*, is **1322.3 cm³**.

9. TRANSFORMATION GEOMETRY

A geometric **transformation** changes the position, size, or shape of a figure. Under a transformation, each point of the original figure is mapped onto its **image**. The original point is called the **preimage**.

9.1 TYPES OF TRANSFORMATIONS

There are four basic types of transformations.

- A **reflection** is a transformation that "flips" a figure over the line of reflection, as shown in the accompanying diagram, so that each figure is the mirror image of the other.

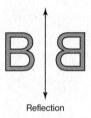

Reflection

Point A' is the reflection of point A in point P if P is the midpoint of $\overline{AA'}$:

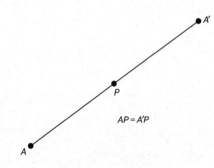

$$AP = A'P$$

- A **translation** is a transformation that "slides" a figure up or down, slides it sideways, or does both, as shown in the accompanying diagram.

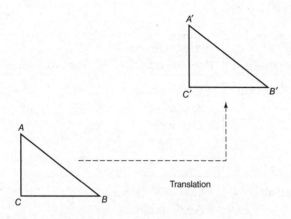

Translation

- A **rotation** is a transformation that "turns" a figure with respect to a fixed reference point, as shown in the accompanying diagram.

Original Figure   Image after 90°
Counterclockwise
Rotation about *P*

- A **dilation** is a transformation that changes the size but not the shape of a figure, as shown in the dilation of the letter **M** in the accompanying diagram.

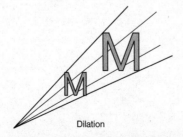

Dilation

9.2 TRANSFORMATIONS USING COORDINATES

The coordinates of the images of points under transformations in the coordinate plane may be determined using the rules in the accompanying table.

Type of Transformation	Properties Preserved	Coordinate Rules
Line reflection	• Collinearity • Angle measure • Distance	• $r_{x\text{-axis}}(x, y) = (x, -y)$ • $r_{y\text{-axis}}(x, y) = (-x, y)$ • $r_{\text{origin}}(x, y) = (-x, -y)$ • $r_{y=x}(x, y) = (y, x)$
Translation	• Collinearity • Angle measure • Distance • Orientation	$T_{h,k}(x,y) = (x + h, y + k)$
Rotation	• Collinearity • Angle measure • Distance • Orientation	• $R_{90°}(x, y) = (-y, x)$ • $R_{180°}(x, y) = (-x, -y)$ • $R_{270°}(x, y) = (y, -x)$
Dilation	• Collinearity • Angle measure • Orientation	$D_k(x,y) = (kx,ky)$, where k is the scale factor

9.3 COMPOSITE TRANSFORMATIONS

A **glide reflection** is a reflection over a line combined with a translation parallel to the reflecting line. The reflection and translation may be performed in either order. In the accompanying figure, △II is the reflection of △II in the x-axis and △III is a translation of △II in a direction that is parallel to the x-axis. Hence, △III is the image of △I under a glide reflection.

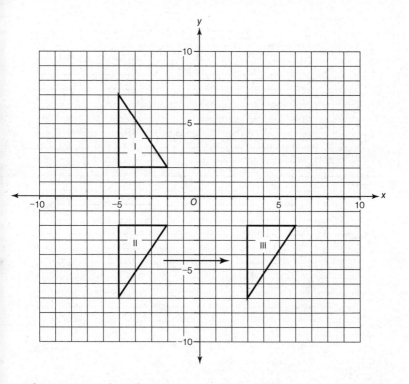

The process of performing a sequence of two or more transformations in which the image of one transformation is used as the preimage of another transformation is called a **composition** of transformations. If T_1 and T_2 represent two transformations, then the notation $T_2 \circ T_1$ represent the composite transformation in which T_1 is performed first and then T_2 is applied to the image of that transformation.

Example 1

In the accompanying figure, △*RST* is the image of △*ABC* under the composition of two different transformations. If one of the transformations is a dilation with center *C*, describe fully the composition of the transformation that maps △*ABC* onto △*RST*.

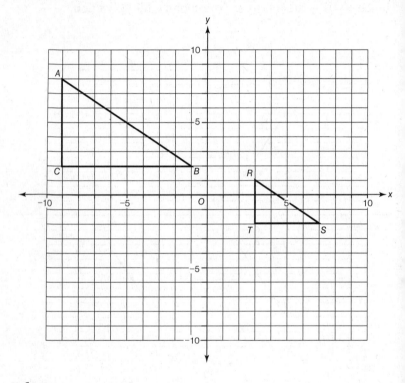

Solution:

- Since each side of △*RST* is one-half the length of the corresponding side of △*ABC*, one of the transformations is a dilation with a scale factor of $\frac{1}{2}$ and center at *C*:

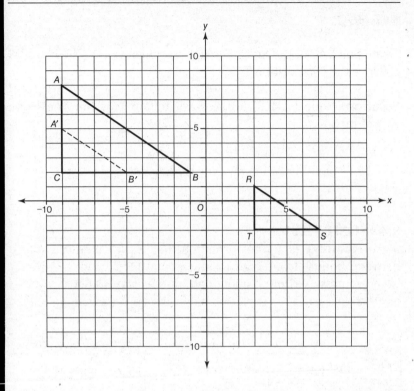

- The transformation that maps $\triangle A'B'C$ onto $\triangle RST$ slides $\triangle A'B'C$ 12 units to the right and 4 units down, which is represented by the translation $T_{12,-4}$.

- Hence, the composite transformation of a dilation with a scale factor of $\frac{1}{2}$ and center at C followed by the translation $T_{12,-4}$, maps $\triangle ABC$ onto $\triangle RST$:

$$T_{12,-4} \circ D_{\frac{1}{2}}(\triangle ABC) \rightarrow \triangle RST.$$

Example 2

Given point $A(-2,3)$. Determine the coordinates of A', the image of A under the composition $T_{-3,4} \circ r_{x\text{-axis}}$.

Solution: Reflect $(-2,3)$ in the x-axis followed by the translation $T_{-3,4}$ of the image:

$$T_{-3,4} \circ r_{x\text{-axis}}A(-2,3) = T_{-3,4}(-2,-3)$$
$$= (-2 - 3, -3 + 4)$$
$$= A'(-5,1)$$

9.4 ISOMETRY

An **isometry** is a transformation that preserves the distance between points. Isometries produce congruent images. A *direct* isometry preserves orientation, and an *opposite* isometry reverses orientation.

Transformation	Isometry?	Type of Isometry
Line reflection	Yes	Opposite
Translation	Yes	Direct
Rotation	Yes	Direct
Dilation	No	
Glide reflection	Yes	Opposite

Practice Exercises

1. Which transformation represents a dilation?

   (1) $(8,4) \rightarrow (11,7)$ (3) $(8,4) \rightarrow (-4,-8)$

   (2) $(8,4) \rightarrow (-8,4)$ (4) $(8,4) \rightarrow (4,2)$

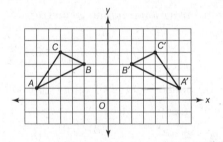

2. In the accompanying diagram, $\triangle A'B'C'$ is the image of $\triangle ABC$. Which type of transformation is shown in the diagram?

   (1) line reflection   (3) translation

   (2) rotation   (4) dilation

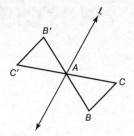

3. Under what type of transformation, shown in the accompanying figure, is $\triangle A'B'C'$ the image of $\triangle ABC$?
(1) dilation
(2) translation
(3) rotation about point A
(4) reflection in line ℓ

4. Which transformation is *not* an isometry?
(1) $T_{5,3}$　　　　　　　　　　(3) $r_{x\text{-axis}}$
(2) D_2　　　　　　　　　　　(4) $R_{O,90°}$

5. What is the image of $P(-4,6)$ under the composite transformation $r_{x=2} \circ r_{y\text{-axis}}$?

(1) $(-8,6)$　　　　　　　　　(3) $(6,0)$
(2) $(4,-2)$　　　　　　　　　(4) $(0,6)$

6. Write an equation of the line of reflection that maps $A(1,5)$ onto $A'(5,1)$.

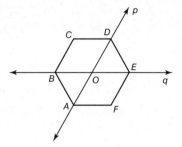

7. In the accompanying figure, p and q are lines of symmetry for regular hexagon $ABCDEF$ intersecting at point O, the center of the hexagon. Determine the image of the composite transformation:

    a. $r_p \circ r_q (\overline{AB})$

    b. $r_q \circ r_p \circ r_q (D)$

8. The coordinates of the vertices of $\triangle ABC$ are $A(2,-3)$, $B(0,4)$, and $C(-1,5)$. If the image of point A under a translation is point $A'(0,0)$, find the images of points B and C under this translation.

9. Graph $\triangle ABC$ with coordinates $A(1,3)$, $B(5,7)$, and $C(8,-3)$. On the same set of axes, graph $\triangle A'B'C'$, the reflection of $\triangle ABC$ in the y-axis.

10. The engineering office in the village of Whitesboro has a map of the village that is laid out on a rectangular coordinate system. A traffic circle located on the map is represented by the equation $(x + 4)^2 + (y - 2)^3 = 81$. The village planning commission asks that the transformation D_2 be applied to produce a new traffic circle where the center of dilation is at the origin.

Find the coordinates of the center of the new traffic circle.

Find the length of the radius of the new traffic circle.

Solutions

1. Under a dilation with a scale factor of k, the image of (x,y) is (kx, ky). Since

$$(8,4) \rightarrow (4,2) = \left(\frac{1}{2} \times 8, \ \frac{1}{2} \times 4 \right),$$

the transformation $(8,4) \rightarrow (4,2)$ represents a dilation with a scale factor of $\frac{1}{2}$.

The correct choice is **(4)**.

2. Because $\triangle A'B'C'$ can be obtained from $\triangle ABC$ by flipping it over the y-axis, the given diagram illustrates a line reflection.

The correct choice is **(1)**.

3. Consider each choice in turn:

(1) A dilation changes the size of the original figure. Since $\triangle ABC$ and $\triangle A'B'C'$ are the same size, the figure is *not* a dilation.

(2) Since $\triangle A'B'C'$ cannot be obtained by sliding $\triangle ABC$ in the horizontal (sideways) or vertical (up-and-down) direction, or in both directions, the transformation is *not* a translation.

(3) A rotation about a fixed point turns a figure about that point. Since angles BAB' and CAC' are straight angles, $\triangle A'B'C'$ is the image of $\triangle ABC$ after a rotation of $180°$ about point A.

(4) Since line ℓ is not the perpendicular bisector of $\overline{BB'}$ and $\overline{CC'}$, points B' and C' are not the images of points B and C, respectively after a reflection in line ℓ.

The correct choice is **(3)**.

4. An *isometry* is a transformation that results in the mapping of a figure onto an image that is congruent to the original figure.

Consider each choice in turn:

(1) $T_{(5,3)}$ is a *translation* which maps a point $P_{(x,y)}$ onto its image, $P'(x + 5, y + 3)$. Since all points are moved 5 units to the right and

3 units up, the distance between any two points is the same as the distance between their respective images. Thus, $T_{(5,3)}$ is an isometry.

(2) D_2 is a *dilation* which maps a point $P(x,y)$ onto its image, $P'(2x,2y)$. The distance between the images of any two points becomes double the distance between the original two points. Thus, D_2 is not an isometry.

(3) $r_{x\text{-axis}}$ is a *reflection* in the x-axis which maps a point $P(x,y)$ onto its image, $P'(x,-y)$. The result is equivalent to flipping figures over the x-axis; all mirror images remain congruent to the original figures. Thus, $r_{x\text{-axis}}$ is an isometry.

(4) $R_{(0,90°)}$ is a *rotation* of 90° about the origin. All original figures remain congruent to their images; they are simply rotated one-fourth of a turn to assume the positions of their images. Thus, $R_{(0,90°)}$ is an isometry.

The correct choice is (**2**).

5. The composite $r_{x=2} \circ r_{y\text{-axis}}$ represents a reflection of a point in the y-axis followed by a reflection of its image in the line $x = 2$.

If P' is the reflection of P in the y-axis, then $PM = P'M$, where $\overline{PP'}$ is perpendicular to the y-axis and M is the intersection of $\overline{PP'}$ with the y-axis. Thus, a reflection in the y-axis replaces a point $P(x,y)$ by its image $P'(-x,y)$. Therefore, $r_{y\text{-axis}}$ $P(-4, 6) \rightarrow P'(4,6)$.

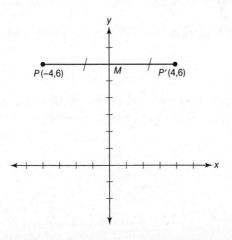

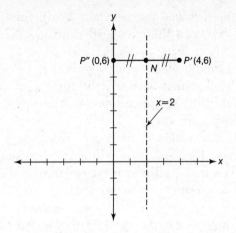

$P'(4,6)$ is next reflected in the line $x = 2$. The line $x = 2$ is a vertical line 2 units to the right of the y-axis. If P'' is the reflection of P' in the line $x = 2$, then $\overline{P'N} = \overline{P''N}$, where $\overline{P'P''}$ is perpendicular to $x = 2$ and N is the intersection of $\overline{P'P''}$ with $x = 2$.

Therefore, a reflection in the line $x = 2$ replaces a point $P'(x,y)$ by its image, $P''(x - 4,y)$. Therefore,

$$r_{x=2}P'(4,6) \rightarrow P''(4 - 4,6) \text{ or } P''(0,6).$$

Combining the results, $r_{x=2} \cdot r_{y\text{-axis}}P(-4,6) \rightarrow P''(0,6)$

The correct choice is (**4**).

6. If $A(1,5)$ is mapped onto $A'(5,1)$, this mapping replaces x by y and y by x:

$$A(x,y) \rightarrow A'(y,x)$$

Such a mapping is produced by a reflection in a line through the origin inclined at $45°$ to the positive directions of both the x- and y-axes.

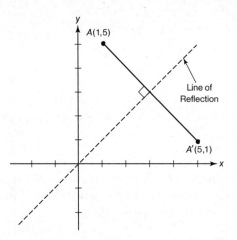

An equation of this line of reflection is $y = x$.

The line of reflection is $\boldsymbol{y = x}$.

7a. Reflect $\overline{AB}$ in line q, followed by the reflection of its image in line p;

$$\overline{AB} \xrightarrow{r_q} \overline{CB} \xrightarrow{r_p} \overline{EF}.$$

b. When the composite is evaluated from right to left, point D is reflected in line q, followed by a reflection in line p, followed by a reflection in line q:

$$D \xrightarrow{r_q} F \xrightarrow{r_p} B \xrightarrow{r_q} B.$$

8. In general, after a translation of h units in the horizontal direction and k units in the vertical direciton, the image of $P'(x,y)$ is $P(x + h, y + k)$. Since

$$A(2,-3) \rightarrow A'(\overbrace{2 + h}, \underbrace{-3 + k}) = A'(0,0),$$

it follows that

$$2 + h = 0 \quad \text{and} \quad h = -2,$$
$$-3 + k = 0 \quad \text{and} \quad k = 3.$$

Therefore:

$$B(0,4) \rightarrow B'(0 + [-2], 4 + 3) = B'(-2,7)$$
$$C(-1,5) \rightarrow C'(-1 + [-2], 5 + 3) = C'(-3,8)$$

B(-2,7), C'(-3,8)

9. After graphing $\triangle ABC$, reflect points A, B, and C in the y-axis, as shown in the accompanying figure. Then connect the image points A', B', and C' with line segments.

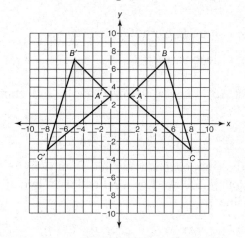

10. The notation D_2 represents a dilation with a scale factor of 2. Under a transformation with a scale factor of 2, the center of the circle remains the same and its radius is multiplied by 2. Since the given equation $(x + 4)^2 + (y - 2)^2 = 81$ can be rewritten as

$$(x - (-4))^2 + (y - 2)^2 = 9^2,$$

the center of the original circle is at $(-4,2)$. Hence, the coordinates of the new traffic circle are also **(-4,2)**.

Because the radius of the original circle is 9, the length of the radius of the new traffic circle is $2 \times 9 = \mathbf{18}$.

10. LOGICAL REASONING AND PROOF

A statement in mathematics is a sentence that can be judged as true or false but not both. If p represents a statement, then the negation of p is the statement with the opposite truth value, denoted by $\sim p$. If p represents "It is raining," then $\sim p$ is the statement, "It is *not* raining."

10.1 TYPES OF LOGICAL CONNECTIVES

Two simple statements, sometimes represented by p and q, can be joined to together to form a new statement as summarized in the accompanying table.

Type of Statement	Symbol Form	Truth Value
Conjunction (AND)	$p \wedge q$	True when p *and* q are both true
Disjunction (OR)	$p \vee q$	True when p is true or q is true or both are true
Conditional (If-then)	$p \rightarrow q$	Always true except when p is true and q is false
Biconditional (If and only if)	$p \leftrightarrow q$	True when p and q have the same truth values

10.2 FORMING RELATED CONDITIONAL STATEMENTS

Starting with the conditional statement "If p, then q," related conditionals can be formed as shown in the accompanying table.

Related Conditional	How to Form Starting with "If p, then q"	Sentence Form
Converse	Interchange p and q.	If q, then p.
Inverse	Negate both p and q.	If not p, then not q.
Contrapositive	Interchange and negate p and q.	If not q, then not p.
Biconditional	Connect p and q with "if and only if"	p if and only if q.

Logically equivalent statements are pairs of statements that always have the same truth value. A conditional statement and its contrapositive are logically equivalent, as are the converse and the inverse of a conditional statement.

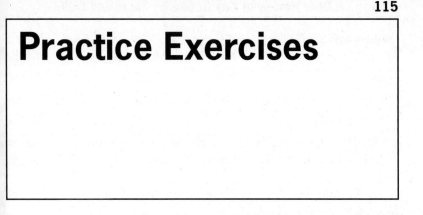

Practice Exercises

1. Given the *false* statement, "Ben has a driver's license or Ben is not 18 years old," what is the truth value of the statement, "Ben is 18 years old"?

2. Given the true statement "I will buy a new suit or I will not go to the dance" and the false statement "I will buy a new suit," what is the truth value of the statement, "I will go the dance"?

3. The statement "Maya plays on the basketball team or Maya joins the ski club" *is false.* Which statement is true?
   (1) Maya plays on the basketball team and Maya joins the ski club.
   (2) Maya plays on the basketball team and Maya does not join the ski club.
   (3) Maya does not play on the basketball team and Maya joins the ski club.
   (4) Maya does not play on the basketball team and Maya does not join the ski club.

4. Which statement is logically equivalent to the statement "If you are an elephant, then you do not forget"?
   (1) If you do not forget, then you are an elephant.
   (2) If you do not forget, then you are not an elephant.
   (3) If you are an elephant, then you forget.
   (4) If you forget, then you are not an elephant.

Solutions

1. Because the given disjunction is false, both disjuncts must be false.

- Hence, the truth value of the statement "Ben is not 18 years old" is false.

- Since a statement and its negation have opposite truth values, the truth value of the statement "Ben is 18 years old" is **true**.

2. Because the given disjunction is true, at least one of the two disjuncts must also be true.

- It is given that the disjunct "I will buy a new suit" is false. This means that the disjunct "I will not go to the dance" is true.

- Since a statement and its negation have opposite truth values, the truth value of the statement "I will go to the dance" is **false**.

3. Because it is given that the statement "Maya plays on the basketball team or Maya joins the ski club" is false, each disjunct must be false. Therefore, the negation of each disjunct must be true. Thus, the statements "Maya does not play on the basketball team" and "Maya does not join the ski club" are both true, so that their conjunction (and) must be true.

The correct choice is **(4)**.

4. A conditional statement and its contrapositive are logically equivalent. To form the contrapositive of a conditional, switch and then negate the hypothesis and conclusion:

- Given conditional: "If you are an elephant, then you do not forget."

- Contrapositive: "If you forget, then you are not an elephant."

The correct choice is **(4)**.

10.3 PROVING TRIANGLES CONGRUENT

Two polygons with the same number of sides are **congruent** if all of their corresponding angles are congruent and all of their corresponding sides are congruent.

The symbol for congruence is ≅.

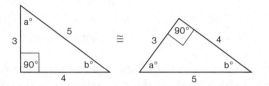

Congruent polygons have the same shape *and* size. To *prove* two *triangles* are congruent, it is sufficient to show that any one of the following conditions is true:

- The three sides of one triangle are congruent to the corresponding parts of the other triangle (SSS ≅ SSS):

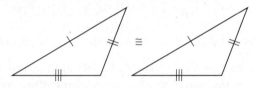

- Two sides and the included angle of one triangle are congruent to the corresponding parts of the other triangle (SAS ≅ SAS):

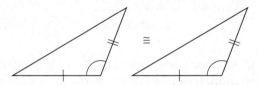

- Two angles and the included side of one triangle are congruent to the corresponding parts of the other triangle (ASA ≅ ASA):

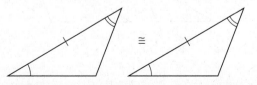

- Two angles and the side opposite one of these angles of one triangle are congruent to the corresponding parts of the other triangle (AAS ≅ AAS):

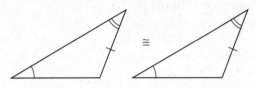

- The hypotenuse and a leg of one right triangle are congruent to the corresponding parts of the other right triangle (HL ≅ HL).

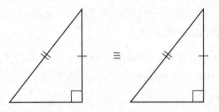

You may *not* conclude that two triangles are congruent when:

- Two sides and an angle that is *not* included of one triangle are congruent to the corresponding parts of the other triangle.

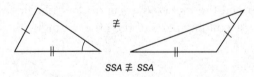

SSA ≢ SSA

- Three angles of one triangle are congruent to the corresponding parts of the other triangle.

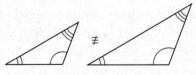

AAA ≢ AAA

After proving two triangles are congruent, you can conclude that any pair of corresponding parts not used in the proof are congruent.

Example

The accompanying diagram shows quadrilateral *BRON* with diagonals $\overline{NR}$ and $\overline{BO}$, which bisect each other at *X*.

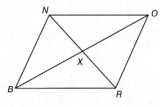

Prove: $\overline{BN} \cong \overline{OR}$.

Solution: Mark off the diagram with the given. Also, mark off the congruent vertical angles. Prove $\triangle BNX \cong \triangle ORX$ by SAS $\cong$ SAS, so $\overline{BN} \cong \overline{OR}$.

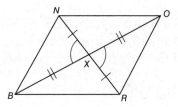

PROOF

Statement	Reason
1. Diagonals $\overline{NR}$ and $\overline{BO}$ bisect each other at X.	1. Given.
2. $\overline{XN} \cong \overline{XR}$. *Side*	2. A bisector of a segment divides it into two congruent segments.
3. $\angle NXB \cong \angle RXO$. *Angle*	3. Vertical angles are congruent.
4. $\overline{XB} \cong \overline{XO}$. *Side*	4. Same as reason 2.
5. $\triangle BNX \cong \triangle ORX$.	5. SAS $\cong$ SAS.
6. $\overline{BN} \cong \overline{OR}$.	6. Corresponding parts of congruent triangles are congruent (CPCTC).

10.4 PROVING TRIANGLES ARE SIMILAR

Two polygons are **similar** if all of their corresponding angles are congruent and the lengths of their corresponding sides are in proportion.

The symbol for similarity is ~.

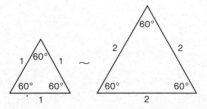

Similar polygons have the same shape but not necessarily the same size. To *prove* two triangles are similar, it is sufficient to show that any one of the following conditions is true:

- Two angles of one triangle are congruent to two angles of the other triangle (AA similarity postulate).

- The lengths of corresponding sides of the two triangles are in proportion (SSS similarity theorem).

- The lengths of two pairs of sides of the two triangles are in proportion, and their included angles are congruent (SAS similarity theorem).

After showing that two triangles are similar, you can conclude that the lengths of pairs of corresponding sides are in proportion.

Example

In the accompanying diagram, $\overline{WA} \parallel \overline{CH}$ and $\overline{WH}$ and $\overline{AC}$ intersect at point T. Prove that $(WT)(CT) = (HT)(AT)$.

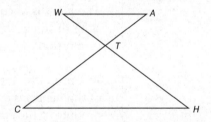

Solution: Reason backward from the Prove. In order to prove $(WT)(CT) = (HT)(AT)$, show that the proportion $\dfrac{WT}{HT} = \dfrac{AT}{CT}$ is true, which first requires proving $\triangle WTA \sim \triangle HTC$.

PROOF

Statement	Reason
1. $\overline{WA} \parallel \overline{CH}$.	1. Given.
2. $\angle W \cong \angle H$. *Angle*	2. If two parallel lines are cut by a transversal, then alternate interior angles are congruent.
3. $\angle A \cong \angle C$. *Angle*	3. Same as reason 2.
4. $\triangle WTA \sim \triangle HTC$.	4. Two triangles are similar if two angles of one triangle are congruent to two angles of the other triangle.
5. $\dfrac{WT}{HT} = \dfrac{AT}{CT}$.	5. The lengths of corresponding sides of similar triangles are in proportion.
6. $(WT)(CT) = (HT)(AT)$.	6. In a proportion, the product of the means is equal to the product of the extremes.

When two triangles overlap a circle, you may be able to show that corresponding pairs of angles of the two triangles are congruent by using the following facts:

- If two angles are measured by the same or congruent arcs, they are congruent.

- An angle inscribed in a semicircle is a right angle.

Example

Given: $\overline{AC}$ is tangent to circle O at A, $\overline{BD}$ is a diameter, chord $\overline{AB}$ is drawn, and $\overline{BC} \perp \overline{AC}$.

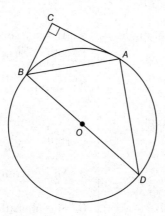

Prove: $\dfrac{BD}{AB} = \dfrac{AB}{BC}$

Solution: To prove the proportion, first prove $\triangle DBA \sim \triangle ABC$ by showing two pairs of corresponding angles are congruent.

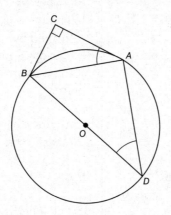

PROOF

Statement	Reason
1. $\overline{BD}$ is a diameter; chord $\overline{AB}$ is drawn.	1. Given.
2. $\angle BAD$ is a right angle.	2. An angle inscribed in a semicircle is a right angle.
3. $\overline{BC} \perp \overline{AC}$.	3. Given.
4. $\angle C$ is a right angle.	4. Perpendicular lines meet at right angles.
5. $\angle BAD \cong \angle C$.	5. All right angles are congruent.
6. $m\angle D = \frac{1}{2}m\overparen{AB}$.	6. The measure of an inscribed angle is equal to one-half the measure of its intercepted arc.
7. $\overline{AC}$ is tangent to circle O at A.	7. Given.
8. $m\angle CAB = \frac{1}{2}m\overparen{AB}$.	8. The measure of an angle formed by a tangent and a chord is equal to one-half of the measure of its intercepted arc.
9. $\angle D \cong \angle CAB$.	9. Angles having the same measure are congruent.
10. $\triangle DBA \sim \triangle ABC$.	10. AA $\cong$ AA.
11. $\dfrac{BD}{AB} = \dfrac{AB}{BC}$.	11. The lengths of corresponding sides of similar triangles are in proportion.

10.5 COORDINATE PROOFS

A coordinate proof may involve applying one or more of the following relationships to figures with variable or numerical coordinates.

- To prove that a quadrilateral is a parallelogram, use the midpoint formula to show that the diagonals have the same midpoint and, as a result, bisect each other.

- To prove that a quadrilateral is a rhombus, use the distance formula to show that the four sides have the same length.

- To prove that two lines are perpendicular, use the slope formula to show that the product of the slopes of the two lines is −1.

- To prove that a quadrilateral is a rectangle, show that it is a parallelogram and that it contains a right angle or that the two diagonals are congruent.

- To prove that a quadrilateral is a trapezoid, use the slope formula to show that: (1) the slopes of two sides are equal and, as a result, the two sides are parallel; and (2) the slopes of two sides are unequal and, as a result, the two sides are not parallel.

Example 1

The coordinates of quadrilateral $PRAT$ are $P(a,b)$, $R(a,b + 3)$, $A(a + 3,b + 4)$, and $T(a + 6,b + 2)$. Prove that $\overline{RA}$ is parallel to $\overline{PT}$.

Solution: To prove that $\overline{RA}$ is parallel to $\overline{PT}$, show that they have the same slope.

- Slope of $\overline{RA}$: $\dfrac{\Delta y}{\Delta x} = \dfrac{(b+4)-(b+3)}{(a+3)-a} = \dfrac{4-3}{3} = \dfrac{1}{3}$

- Slope of $\overline{PT}$: $\dfrac{\Delta y}{\Delta x} = \dfrac{(b+2)-(b)}{(a+6)-a} = \dfrac{2}{6} = \dfrac{1}{3}$

Hence, $\overline{RA} \parallel \overline{PT}$.

Example 2

Quadrilateral *FAME* has vertices $F(2,-2)$, $A(8, -1)$, $M(9,3)$, and $E(3,2)$. Prove that *FAME* is a parallelogram but *not* a rectangle.

Solution: First, prove that *FAME* is a parallelogram. Then show either that its diagonals are not congruent or that a pair of adjacent sides do not form a right angle.

<u>Method 1</u>: Use the midpoint formula to show that the quadrilateral is a parallelogram, and the distance formula to compare the lengths of the diagonals.

A quadrilateral is a parallelogram if its diagonals have the same midpoint.

- The coordinates of the midpoint of diagonal $\overline{FM}$ are

$$\left(\frac{2+9}{2},\frac{-2+3}{2}\right)=\left(\frac{11}{2},\frac{1}{2}\right).$$

- The coordinates of the midpoint of diagonal $\overline{AE}$ are

$$\left(\frac{8+3}{2},\frac{-1+2}{2}\right)=\left(\frac{11}{2},\frac{1}{2}\right).$$

- Since the diagonals of quadrilateral *FAME* have the same midpoint, the diagonals bisect each other and, as a result, *FAME* is a parallelogram.

If the diagonals of a parallelogram have the same length, the parallelogram is a rectangle. Use the distance formula to find the lengths of diagonals $\overline{FM}$ and $\overline{AE}$:

$$FM = \sqrt{(9-2)^2+(3-(-2))^2} = \sqrt{7^2+5^2} = \sqrt{74}$$

$$AE = \sqrt{(3-8)^2+(2-(-1))^2} = \sqrt{(-5)^2+3^2} = \sqrt{34}$$

Since $FM \neq AE$, parallelogram *FAME* is *not* a rectangle.

<u>Method 2</u>: Use the slope formula to compare the slopes of opposite and adjacent sides.

A quadrilateral is a parallelogram if its opposite sides have the same slope and, as a result, are parallel.

- Slope $\overline{FA} = \dfrac{\Delta y}{\Delta x} = \dfrac{-1-(-2)}{8-2} = \dfrac{-1+2}{6} = \dfrac{1}{6}$

- Slope $\overline{AM} = \dfrac{\Delta y}{\Delta x} = \dfrac{3-(-1)}{9-8} = \dfrac{3+1}{1} = \dfrac{1}{3}$

- Slope $\overline{ME} = \dfrac{\Delta y}{\Delta x} = \dfrac{2-3}{3-9} = \dfrac{-1}{-6} = \dfrac{1}{6}$

- Slope $\overline{FE} = \dfrac{\Delta y}{\Delta x} = \dfrac{2-(-2)}{3-2} = \dfrac{2+2}{1} = 4$

Since slope $\overline{FA}$ = slope $\overline{ME} = \dfrac{1}{6}$, $\overline{FA} \parallel \overline{ME}$. Also, slope $\overline{AM}$ = slope $\overline{FE}$ = 4, so $\overline{AM} \parallel \overline{FE}$. Hence, *FAME* is a parallelogram because both pairs of opposite sides are parallel.

Because $\dfrac{1}{6}$ and 4 are *not* negative reciprocals, $\overline{FA}$ and $\overline{AM}$ do not form a right angle, so *FAME* is not a rectangle.

10.6 INDIRECT METHOD OF PROOF

To prove a statement indirectly, assume what needs to be proved is *not* true. Then show that this leads to contradiction of a known fact.

Example

Given: $\triangle ABC$ is scalene;
$\overline{BD}$ bisects $\angle ABC$.
Prove: $\overline{BD}$ is *not* perpendicular
to $\overline{AC}$.

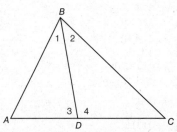

Solution: Assume the opposite of the Prove: $\overline{BD} \perp \overline{AC}$. Then $\angle 3 \cong \angle 4$ since perpendicular lines intersect at right angles and all right angles are congruent. Because $\overline{BD} \cong \overline{BD}$ and $\angle 1 \cong \angle 2$ (definition of an angle bisector), $\triangle ABD \cong \triangle CBD$ by ASA $\cong$ ASA. By CPCTC, $\overline{AB} \cong \overline{BC}$. But this contradicts the Given which states that $\triangle ABC$ is scalene. Hence, the assumption that $\overline{BD} \perp \overline{AC}$ must be false. It must be the case that $\overline{BD}$ is not perpendicular to $\overline{AC}$, as this is the only other possibility. An indirect proof can also be presented using the two-column format:

PLAN: Assume $\overline{BD} \perp \overline{AC}$ and show that this means $\triangle ABD \cong \triangle CDB$. Therefore, $\overline{AB} \cong \overline{BC}$, which contradicts the Given that $\triangle ABC$ is scalene.

PROOF

Statement	Reason
1. Assume $\overline{BD} \perp \overline{AC}$.	1. Assume that the opposite of what needs to be proved is true.
2. $\angle 3 \cong \angle 4$. *Angle*	2. Perpendicular lines interesect to form congruent adjacent angles.
3. $\overline{BD} \cong \overline{BD}$. *Side*	3. Reflexive property of congruence.
4. $\angle 1 \cong \angle 2$. *Angle*	4. It is given that BD bisects $\angle ABC$.
5. $\triangle ADB \cong \triangle CDB$.	5. ASA postulate.
6. $\overline{AB} \cong \overline{BC}$.	6. CPCTC.
7. $\triangle ABC$ is scalene.	7. Given.
8. Statement 7 contradicts statement 6.	8. In a scalene triangle, no two sides have the same length.
9. $\overline{BD}$ is *not* perpendicular to $\overline{AC}$.	9. Because statement 1 leads to a contradiction, its opposite is true.

Practice Exercises

1. The lengths of the five sides of a pentagon are 1, 3, 5, 7, and 12. If the length of the longest side of a similar pentagon is 18, what is the perimeter of the larger pentagon?
   (1) 24   (3) 42
   (2) 36   (4) 48

2. If the midpoints of the sides of a triangle are connected, the area of the triangle formed is what part of the area of the original triangle?

   (1) $\dfrac{1}{4}$ (3) $\dfrac{3}{8}$

   (2) $\dfrac{1}{3}$ (4) $\dfrac{1}{2}$

3. At a certain time during the day, light falls so that a pole 10 feet in height casts a shadow 15 feet in length on level ground. At the same time, a man casts a shadow that is 9 feet in length. How tall is the man?

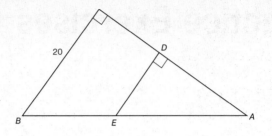

4. In the accompanying diagram of right triangle *ACB*, ∠*C* is a right angle. Point *D* is on leg $\overline{AC}$, and point *E* is on hypotenuse $\overline{AB}$ with $\overline{AD} \perp \overline{DE}$. Explain why △*ABC* is similar to △*AED*. If *BC* = 20 and the ratio of *AD* to *CD* is 3:2, find the length of $\overline{DE}$.

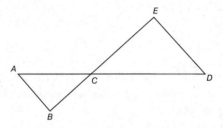

5. In the accompanying diagram, $\overline{AB} \perp \overline{BE}$ and $\overline{DE} \perp \overline{BE}$. Explain why △*ABC* is similar to △*DEC*. If *AC* = 8, *DC* = 12, and *BE* = 15, what is *BC*?

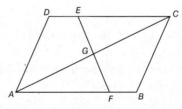

6. In the accompanying diagram of parallelogram *ABCD*, $\overline{DE} \cong \overline{BF}$. Triangle *EGC* can be proved congruent to triangle *FGA* by

(1) HL ≅ HL
(2) AAA ≅ AAA
(3) AAS ≅ AAS
(4) SSA ≅ SSA

7. Given: $\overline{MK} \perp \overline{JL}, \overline{LP} \perp \overline{JM},$
$\overline{JK} \cong \overline{JP}$
Prove: $\triangle JMK \cong \triangle JLP$

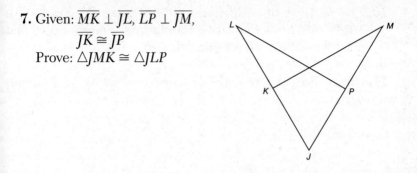

8. Given: Circle O, $\overline{DB}$ is tangent to the circle at B, $\overline{BC}$ and $\overline{BA}$ are chords, and C is the midpoint of $\overset{\frown}{AB}$.

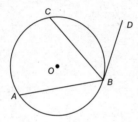

Prove: $\angle ABC \cong \angle CBD$

9. Given: Rectangle $ABCD$, $\overline{BNPC}, \overline{AEP}, \overline{DEN}$, and $\overline{AP} \cong \overline{DN}$.

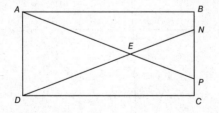

Prove: **a.** $\triangle ABP \cong \triangle DCN$
 b. $\overline{AE} \cong \overline{DE}$

10. The vertices of quadrilateral *GAME* are $G(r,s)$, $A(0,0)$, $M(t,0)$, and $E(t + r,s)$. Using coordinate geometry, prove that quadrilateral *GAME* is a parallelogram.

11. Given: Isosceles triangle ABC, $\overline{BA} \cong \overline{BC}$, $\overline{AE} \perp \overline{BC}$, and $\overline{BD} \perp \overline{AC}$.

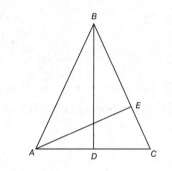

Prove: $\dfrac{AC}{BA} = \dfrac{AE}{BD}$

12. Given: $\overline{AB} \cong \overline{AC}$; $\overline{BD} \not\cong \overline{CD}$.

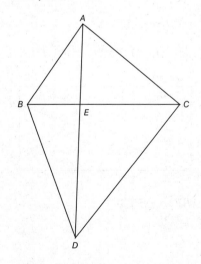

Prove: $\overline{BE} \not\cong \overline{EC}$

13. Given: $\square ABCD$;
    E is the midpoint of $\overline{AD}$;
    F is the midpoint of $\overline{BC}$.

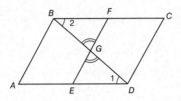

Prove: G is the midpoint of $\overline{EF}$.

14. Given: $\overline{AB} \cong \overline{AC}$; $\overline{BD} \cong \overline{CE}$, $\overline{BF} \perp \overline{AD}$,
    $\overline{CG} \perp \overline{AE}$, and $\angle 1 \cong \angle 2$.

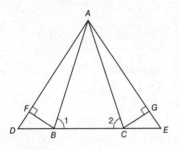

Prove: $\overline{DF} \cong \overline{EG}$

15. In the accompanying diagram of circle O, diameter $\overline{AOB}$ is drawn, chords $\overline{AC}$, $\overline{CD}$, and $\overline{BC}$ are drawn, secant $\overline{EAD}$, and $\overline{EC} \perp \overline{CD}$.

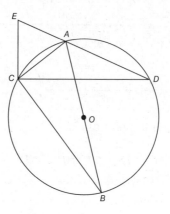

Prove: $\triangle ABC \sim \triangle EDC$

16. Given: $ABCD$ is a parallelogram.
Prove: $KM \times LB = LM \times KD$

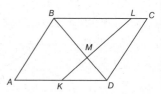

Solutions

1. Let x = perimeter of the larger pentagon. Then:

$$\frac{\text{longest side of smaller pentagon}}{\text{longest side of larger pentagon}} = \frac{\text{perimeter of smaller pentagon}}{\text{perimeter of larger pentagon}}$$

$$\frac{12}{18} = \frac{1+3+5+7+12}{x}$$

$$\frac{2}{3} = \frac{28}{x}$$

$$2x = 84$$

$$x = \frac{84}{2}$$

$$= 42$$

The correct choice is **(3)**.

2. The triangle formed by joining the midpoints of the sides of a triangle is similar to the original triangle such that the ratio of the lengths of their corresponding sides is 1:2. Since the ratio of the areas of similar triangles is the same as the square of the ratio of the lengths of corresponding sides, the area of the triangle formed is $\left(\frac{1}{2}\right)^2$ or $\frac{1}{4}$ of the area of the original triangle.

The correct choice is **(1)**.

3. The shadow cast by the pole and the man can be represented as legs of right triangles in which the hypotenuses represent the rays of light.

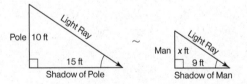

Assume that in each right triangle the ray of light makes the same angle with the ground. Since two angles of the first right triangle are congruent to two angles of the second right triangle, the two right triangles are similar and, as a result, the lengths of their corresponding sides are in proportion. Hence:

$$\frac{\text{height of pole}}{\text{height of man}} = \frac{\text{shadow of pole}}{\text{shadow of man}}$$

$$\frac{10}{x} = \frac{15}{9}$$

$$15x = 90$$

$$x = \frac{90}{15}$$

$$= 6$$

4. Since $\angle C \cong \angle D$ (right angles are congruent) and $\angle A \cong \angle A$, $\triangle ABC \sim \triangle AED$. Because it is given that $AD{:}CD = 3{:}2$, represent the lengths of AD and CD by $3x$ and $2x$, respectively. The lengths of corresponding sides of similar triangles are in proportion. Thus:

$$\frac{DE}{BC} = \frac{AD}{AC}$$

$$\frac{DE}{20} = \frac{3x}{2x + 3x}$$

$$\frac{DE}{20} = \frac{3x}{5x}$$

$$\frac{DE}{20} = \frac{3}{5}$$

$$5(DE) = 60$$

$$\frac{5(DE)}{5} = \frac{60}{5}$$

$$DE = 12$$

The length of $\overline{DE}$ is **12**.

5. Since $\angle B \cong \angle E$ (right angles are congruent) and $\angle ACB \cong \angle DCE$ (vertical angles are congruent), $\triangle ABC \sim \triangle DEC$. The lengths of corresponding sides of similar triangles are in proportion. If x represents the length of $\overline{BC}$, then $15 - x$ represents the length of $\overline{EC}$. Thus:

$$\frac{AC}{DC} = \frac{BC}{EC}$$

$$\frac{8}{12} = \frac{x}{15 - x}$$

$$12x = 8(15 - x)$$

$$12x = 120 - 8x$$

$$12x + 8x = 120$$

$$\frac{20x}{20} = \frac{120}{60}$$

$$x = 6$$

BC = 6.

6. Because vertical angles are congruent, $\angle EGC \cong \angle FGA$ (A). Since opposite sides of a parallelogram are parallel, alternate interior angles are congruent, so $\angle CEG \cong \angle AFG$ (A). Opposite sides of a parallelogram are congruent, and it is given that $\overline{DE} \cong \overline{BF}$. Using the subtraction property, it must be the case that $\overline{CE} \cong \overline{AF}$ (S). Hence, triangle EGC could be proved congruent to triangle FGA by AAS $\cong$ AAS.

The correct choice is **(3)**.

7. PLAN: Since $\angle J \cong \angle J$ (A), $JK \cong JP$ (S), and $\angle 1 \cong \angle 2$ (A), prove the triangles congruent by using the ASA postulate.

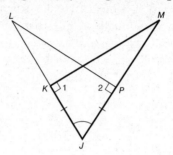

PROOF

Statement	Reason
1. $\overline{MK} \perp \overline{JL}$, $\overline{LP} \perp \overline{JM}$.	1. Given.
2. Angles 1 and 2 are right angles.	2. Perpendicular lines interesect to form right angles.
3. $\angle 1 \cong \angle 2$. *Angle*	3. All right angles are congruent.
4. $\overline{JK} \cong \overline{JP}$. *Side*	4. Given.
5. $\angle J \cong \angle J$. *Angle*	5. Reflexive property of congruence.
6. $\triangle JMK \cong \triangle JLP$.	6. ASA postulate.

8. Given: Circle O, $\overline{DB}$ is tangent to the circle at B, $\overline{BC}$ and $\overline{BA}$ are chords, and C is the midpoint of $\overset{\frown}{AB}$.

Prove: $\angle ABC \cong \angle CBD$

PLAN: Show that the measures of angles ABC and CBD are one-half the measures of equal arcs and, as a result, the angles are congruent.

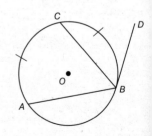

PROOF

Statement	Reason
1. C is the midpoint of $\overset{\frown}{AB}$.	1. Given.
2. $m\overset{\frown}{AC} = m\overset{\frown}{BC}$.	2. The midpoint of an arc divides the arc into two arcs that have the same measure.
3. $m\angle ABC = \frac{1}{2}m\overset{\frown}{AC}$.	3. The measure of an inscribed angle is one-half the measure of its intercepted arc.
4. $\overline{DB}$ is tangent to the circle O at B.	4. Given.
5. $m\angle CBD = \frac{1}{2}m\overset{\frown}{DC}$.	5. The measure of an angle formed by a tangent and a chord is one-half the measure of its intercepted arc.
6. $m\angle ABC \cong m\angle CBD$.	6. Halves of equal quantities are equal.
7. $\angle ABC \cong \angle CBD$.	7. Angles that are equal in measure are congruent.

9. Given: Rectangle $ABCD$, $\overline{BNPC}$, $\overline{AEP}$, $\overline{DEN}$, and $\overline{AP} \cong \overline{DN}$.
   Prove: **a.** $\triangle ABP \cong \triangle DCN$
          b. $\overline{AE} \cong \overline{DE}$

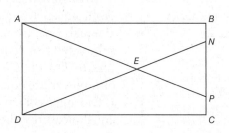

PLAN: **a.** Prove right triangle $ABP \cong$ right triangle DCN by
HL $\cong$ HL.

b. Prove $\overline{AE} \cong \overline{DE}$ by showing that the angles opposite
these sides in $\triangle AED$ are congruent:

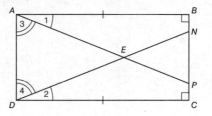

PROOF

Statement	Reason
a. Prove $\triangle ABP \cong \triangle DCN$.	
1. $ABCD$ is a rectangle.	1. Given.
2. $\angle A$, $\angle B$, $\angle C$, and $\angle D$ are right angles.	2. A rectangle contains four right angles.
3. $\triangle ABP$ and $\triangle DCN$ are right triangles.	3. A triangle that contains a right angle is a right triangle.
4. $\overline{AP} \cong \overline{DN}$ (H).	4. Given.
5. $\overline{AB} \cong \overline{CD}$ (L).	5. Opposite sides of a rectangle are congruent.
6. $\triangle ABP \cong \triangle DCN$.	6. HL $\cong$ HL.
b. Prove $\overline{AE} \cong \overline{DE}$.	
7. $m\angle 1 = m\angle 2$.	7. Corresponding angles of congruent triangles are equal in measure.
8. $m\angle BAD = m\angle CDA$.	8. Right angles are equal in measure.
9. $m\angle 3 = m\angle 4$.	9. If equals ($m\angle 1 = m\angle 2$) are subtracted from equals ($m\angle BAD = m\angle CDA$), the differences are equal.
10. $\overline{AE} \cong \overline{DE}$.	10. If two angles of a triangle have equal measures, the sides opposite these angles are congruent.

10. A quadrilateral is a parallelogram if its diagonals have the same midpoint and, as a result, bisect each other.

- Determine the x- and y-coordinates of the midpoint of diagonal $\overline{GM}$ by finding the averages of the corresponding coordinates of $G(r,s)$ and $M(t,0)$:

$$\text{Midpoint of } \overline{GM} = \left(\frac{r+t}{2}, \frac{s+0}{2} \right) = \left(\frac{r+t}{2}, \frac{s}{2} \right)$$

- Determine the x- and y-coordinates of the midpoint of diagonal $\overline{AE}$ by finding the averages of the corresponding coordinates of $A(0,0)$ and $E(t+r,s)$:

$$\text{Midpoint of } \overline{AE} = \left[\frac{0+(t+r)}{2}, \frac{0+s}{2} \right] = \left(\frac{r+t}{2}, \frac{s}{2} \right)$$

- Diagonals $\overline{GM}$ and $\overline{AE}$ of quadrilateral $GAME$ have the same midpoint; hence they bisect each other.

Therefore, quadrilateral $GAME$ is a parallelogram.

11. Given: Isosceles triangle ABC, $\overline{BA} \cong \overline{BC}$, $\overline{AE} \perp \overline{BC}$, and $\overline{BD} \perp \overline{AC}$.

Prove: $\dfrac{AC}{BA} = \dfrac{AE}{BD}$

PLAN: Show that triangles AEC and BDA are similar and that, as a result, the lengths of their corresponding sides are in proportion.

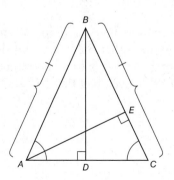

PROOF

Statement	Reason
1. $\overline{BA} \cong \overline{BC}$.	1. Given.
2. $\angle BAD \cong \angle ACE$ (A).	2. If two sides of a triangle are congruent, then the angles opposite these sides are congruent.
3. $\overline{AE} \perp \overline{BC}$, $\overline{BD} \perp \overline{AC}$.	3. Given.
4. Angles BDA and AEC are right angles.	4. Perpendicular lines meet to form right angles.
5. $\angle BDA \cong \angle AEC$ (A).	5. All right angles are congruent.
6. $\triangle AEC \sim \triangle BDA$.	6. Two triangles are similar if two angles of one triangle are congruent to two angles of the other triangle.
7. Prove: $\dfrac{AC}{BA} = \dfrac{AE}{BD}$.	7. The lengths of corresponding sides of similar triangles are in proportion.

12. Given: $\overline{AB} \cong \overline{AC}$;
$\quad\quad\quad\overline{BD} \not\cong \overline{CD}$.
Prove: $\overline{BE} \not\cong \overline{EC}$

PLAN: Use an indirect proof. Assume that $\overline{BE} \cong \overline{EC}$ and show that this leads to a contradiction of the given fact that $\overline{BD} \not\cong \overline{CD}$.

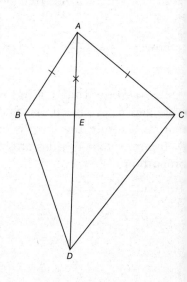

PROOF

Statement	Reason
1. $\overline{AB} \cong \overline{AC}$ (S).	1. Given.
2. $\overline{BD} \not\cong \overline{CD}$.	2. Given.
3. Either $\overline{BE} \not\cong \overline{EC}$ or $\overline{BE} \cong \overline{EC}$.	3. A statement is either true or false.
4. $\overline{BE} \cong \overline{EC}$ (S).	4. Assume that the opposite of what needs to be proved is true.
5. $\overline{AE} \cong \overline{AE}$ (S).	5. Reflective property of congruence.
6. $\triangle AEB \cong \triangle AEC$.	6. SSS $\cong$ SSS.
7. $\angle AEB \cong \angle AEC$.	7. Corresponding parts of congruent triangles are congruent.
8. $\angle BED \cong \angle CED$ (A).	8. Supplements of congruent angles are congruent.
9. $\overline{ED} \cong \overline{ED}$ (S).	9. Reflexive property of congruence.
10. $\triangle BED \cong \triangle CED$.	10. SAS $\cong$ SAS.
11. $\overline{BD} \cong \overline{CD}$.	11. Corresponding parts of congruent triangles are congruent.
12. $\overline{BE} \not\cong \overline{EC}$	12. Statement 11 contradicts statement 2. The assumption made in statement 4 must be false, and so its opposite must be true.

13. PLAN: Show that $\triangle DGE \cong \triangle BGF$ by the AAS Theorem.

PROOF

Statement	Reason
1. $\square ABCD$.	1. Given.
2. $\angle DGE \cong \angle BGF$. *Angle*	2. Vertical angles are congruent.
3. $\overline{AD} \parallel \overline{BC}$.	3. Opposite sides of a parallelogram are parallel.
4. $\angle 1 \cong \angle 2$. *Angle*	4. If two lines are parallel, then alternate interior angles are congruent.
5. $AE = BC$.	5. Opposite sides of a parallelogram are equal in length.
6. E is the midpoint of $\overline{AD}$; F is the midpoint of $\overline{BC}$.	6. Given.
7. $DE = \frac{1}{2}AD$, $BF = \frac{1}{2}BC$.	7. Definition of midpoint.
8. $DE = BF$.	8. Halves of equals are equal.
9. $\overline{DE} \cong \overline{BF}$. *Side*	9. Segments having equal lengths are congruent.
10. $\triangle DGE \cong \triangle BGF$.	10. AAS theorem.
11. $\overline{EG} \cong \overline{FG}$.	11. CPCTC.
12. G is the midpoint of $\overline{EF}$.	12. Definition of midpoint.

14. Write a paragraph proof by reasoning as follows:

To prove $\overline{DF} \cong \overline{EG}$, we must prove $\triangle DFB \cong \triangle EGC$. Before we can prove these triangles congruent, we need to show that $\triangle ABD \cong \triangle ACE$. These triangles are congruent by SAS $\cong$ SAS since $\overline{AB} \cong \overline{AC}$ (Given); $\angle ABD \cong \angle ACE$ (supplements of congruent angles 1 and 2 are congruent); $\overline{BD} \cong \overline{CE}$ (Given). Because $\triangle ABD \cong \triangle ACE$, $\angle D \cong \angle E$. We can use these congruent angles to prove that $\triangle DFB \cong \triangle EGC$. We know that $\angle D \cong \angle E$ (CPCTC), $\angle DFB \cong \angle EGC$ (right angles are congruent), and $\overline{BD} \cong \overline{CE}$. Therefore, $\triangle DFB \cong \triangle EGC$ by AAS $\cong$ AAS. Hence, $\overline{DF} \cong \overline{EG}$ by CPCTC.

15. Prove $\triangle ABC \sim \triangle EDC$ by showing that two pairs of corresponding angles are congruent.

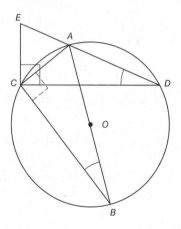

PROOF

Statement	Reason
1. $\overline{EC} \perp \overline{CD}$.	1. Given.
2. $\angle ECD$ is a right angle.	2. Perpendicular lines intersect to form right angles.
3. Angle $\angle ACB$ is a right angle.	3. An angle inscribed in a semicircle is a right angle.
4. $\angle ACB \cong \angle ECD$.	4. All right angles are congruent.
5. $\angle B \cong \angle D$.	5. In a circle, inscribed angles that intercept the same arc are congruent.
6. $\triangle ABC \sim \triangle EDC$.	6. Two triangles are similar if two pairs of corresponding angles are congruent.

16. Prove $\triangle KMD \sim \triangle LMB$ by showing that two pairs of corresponding angles are congruent.

PROOF

Statement	Reason
1. $\square ABCD$.	1. Given.
2. $\overline{AD} \parallel \overline{BC}$.	2. Opposite sides of a parallelogram are parallel.
3. $\angle 1 \cong \angle 2$; $\angle 3 \cong \angle 4$.	3. If two lines are parallel, then their alternate interior angles are congruent.
4. $\triangle KMD \sim \triangle LMB$.	4. AA theorem.
5. $\dfrac{KM}{LM} = \dfrac{KD}{LB}$.	5. The lengths of corresponding sides of similar triangles are in proportion.
6. $KM \times LB = LM \times KD$.	6. In a proportion, the product of the means equals the product of the extremes.

Geometric Constructions

BASIC CONSTRUCTIONS

Geometric constructions, unlike drawings, are made with only a straightedge and a compass. The point at which the sharp point of the compass is placed is called the **center**, and the fixed compass setting that is used to draw **arcs** is the **radius length**. Here are two basic constructions.

- To construct a segment that is congruent to a given segment:

Step	Diagram
1. Draw any line and choose any convenient point on it. Label the line as ℓ and the point as C.	
2. Using a compass, measure $\overline{AB}$ by placing the compass point on A and the pencil point on B.	
3. Using the same compass setting, place the compass point on C and draw an arc that interesects line ℓ. Label the point of intersection as D.	

Conclusion: $\overline{AB} \cong \overline{CD}$.

• To construct an angle that is congruent to a given angle:

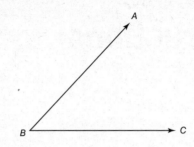

Step	Diagram
1. Draw any line and choose any convenient point on it. Label the line as ℓ and the point as S.	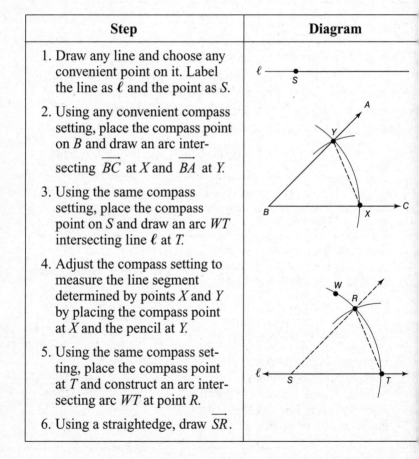
2. Using any convenient compass setting, place the compass point on B and draw an arc intersecting $\overrightarrow{BC}$ at X and $\overrightarrow{BA}$ at Y.	
3. Using the same compass setting, place the compass point on S and draw an arc WT intersecting line ℓ at T.	
4. Adjust the compass setting to measure the line segment determined by points X and Y by placing the compass point at X and the pencil at Y.	
5. Using the same compass setting, place the compass point at T and construct an arc intersecting arc WT at point R.	
6. Using a straightedge, draw $\overrightarrow{SR}$.	

Conclusion: $\angle ABC \cong \angle RST$.

Reason: The arcs were constructed so that $\overline{BX} \cong \overline{ST}$, $\overline{BY} \cong \overline{SR}$, and $\overline{XY} \cong \overline{TR}$. Therefore, $\triangle XYB \cong \triangle TRS$ by the SSS postulate. Since all of the corresponding pairs of parts of congruent triangles are congruent, $\angle ABC \cong \angle RST$.

REQUIRED CONSTRUCTIONS

There are six constructions that you are required to know.

Required Construction 1: To construct an equilateral triangle using a given segment length.

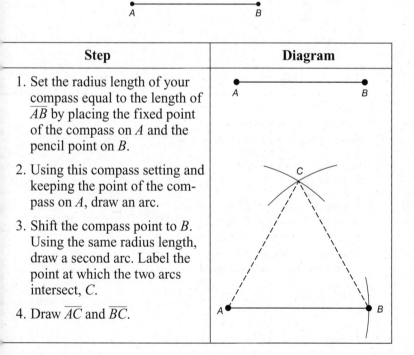

Step	Diagram
1. Set the radius length of your compass equal to the length of $\overline{AB}$ by placing the fixed point of the compass on A and the pencil point on B.	
2. Using this compass setting and keeping the point of the compass on A, draw an arc.	
3. Shift the compass point to B. Using the same radius length, draw a second arc. Label the point at which the two arcs intersect, C.	
4. Draw $\overline{AC}$ and $\overline{BC}$.	

Conclusion: $\triangle ABC$ is equilateral.

Reason: The distance from A to any point on the first arc is AB, and the distance from B to any point on the second arc is also AB. Therefore, the third vertex of the equilateral triangle is the point at which these two arcs intersect, labeled point C, since $AC = BC = AB$.

Required Construction 2: To construct the bisector of a given angle..

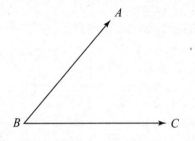

Step	Diagram
1. Using B as a center, construct an arc, using any convenient radius length, that intersects $\overrightarrow{BA}$ at point P and $\overrightarrow{BC}$ at point Q. 2. Using points P and Q as centers and the same radius length, draw a pair of arcs that intersect. Label the point at which the arcs intersect as D. 3. Draw $\overrightarrow{BD}$.	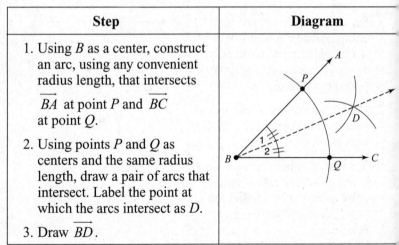

Conclusion: $\overrightarrow{BD}$ is the bisector of $\angle ABC$.

Reason: The arcs were constructed so that $BP = BQ$ and $PD = QD$, thus making $\triangle BPD \cong \triangle BQD$. Therefore, m$\angle 1 =$ m$\angle 2$.

Required Construction 3: To construct the perpendicular bisector of a given segment.

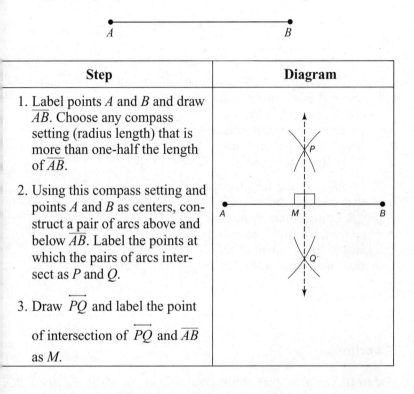

Step	Diagram
1. Label points A and B and draw $\overline{AB}$. Choose any compass setting (radius length) that is more than one-half the length of $\overline{AB}$. 2. Using this compass setting and points A and B as centers, construct a pair of arcs above and below $\overline{AB}$. Label the points at which the pairs of arcs intersect as P and Q. 3. Draw $\overleftrightarrow{PQ}$ and label the point of intersection of $\overleftrightarrow{PQ}$ and $\overline{AB}$ as M.	

Conclusion: $\overleftrightarrow{PQ}$ is the perpendicular bisector of $\overline{AB}$.

Reason: The arcs were constructed in such a way that $AP = BP = AQ = BQ$. Since quadrilateral $APBQ$ is equilateral, it is a rhombus. Since the diagonals of a rhombus are perpendicular bisectors, $\overline{AM} \cong \overline{BM}$ and $\overleftrightarrow{PQ} \perp \overline{AB}$.

Required Construction 4: To construct a line perpendicular to a given line at a given point on the line.

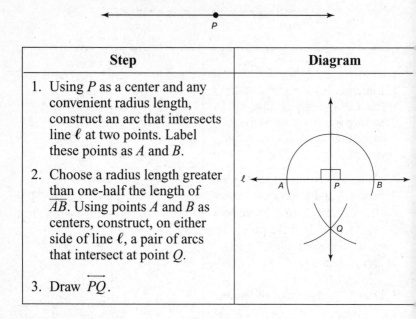

Step	Diagram
1. Using P as a center and any convenient radius length, construct an arc that intersects line ℓ at two points. Label these points as A and B.	
2. Choose a radius length greater than one-half the length of $\overline{AB}$. Using points A and B as centers, construct, on either side of line ℓ, a pair of arcs that intersect at point Q.	
3. Draw $\overleftrightarrow{PQ}$.	

Conclusion: $\overleftrightarrow{PQ} \perp \ell$.

Reason: The arcs were constructed so that $AP = PB$ and $AQ = BQ$, thus making $\triangle APQ \cong \triangle BPQ$ by SSS $\cong$ SSS. This means that angles APQ and BPQ are both congruent and adjacent; so each must be a right angle, making $\overleftrightarrow{PQ} \perp \ell$.

Required Construction 5: To construct a line parallel to a given line through a given point.

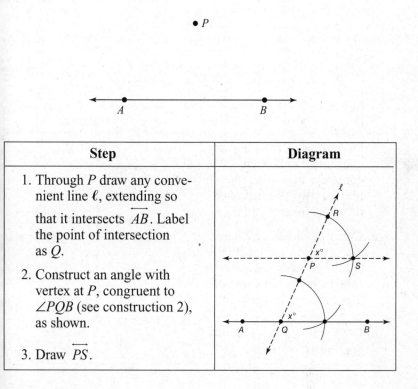

Step	Diagram
1. Through P draw any convenient line ℓ, extending so that it intersects $\overleftrightarrow{AB}$. Label the point of intersection as Q. 2. Construct an angle with vertex at P, congruent to $\angle PQB$ (see construction 2), as shown. 3. Draw $\overleftrightarrow{PS}$.	

Conclusion: $\overleftrightarrow{PS} \parallel \overleftrightarrow{AB}$.

Reason: The arcs were constructed so that $\angle PQB$ and $\angle RPS$ are congruent corresponding angles, making $\overleftrightarrow{PS} \parallel \overleftrightarrow{AB}$.

Required Construction 6: To construct a line perpendicular to a given line through a given point not on the line.

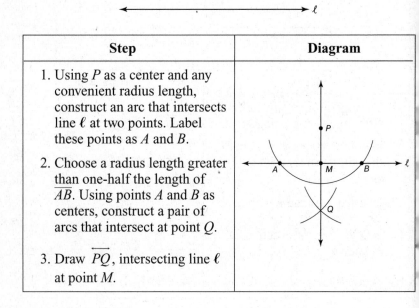

Step	Diagram
1. Using *P* as a center and any convenient radius length, construct an arc that intersects line ℓ at two points. Label these points as *A* and *B*. 2. Choose a radius length greater than one-half the length of $\overline{AB}$. Using points *A* and *B* as centers, construct a pair of arcs that intersect at point *Q*. 3. Draw $\overleftrightarrow{PQ}$, intersecting line ℓ at point *M*.	

Conclusion: $\overleftrightarrow{PQ}$ is perpendicular to ℓ at point *M*.

Glossary of Terms

acute angle An angle whose degree measure is less than 90 and greater than 0.

acute triangle A triangle with three acute angles.

adjacent angles Two angles with the same vertex, a common side, and no interior points in common.

altitude In a triangle, a segment that is perpendicular to the side to which it is drawn.

angle The union of two rays that have the same endpoint.

angle of rotational symmetry The smallest positive angle through which a figure with rotational symmetry can be rotated so that it coincides with itself. For a regular n–polygon, this angle is $\dfrac{360}{n}$.

apothem The radius of the inscribed circle of a regular polygon.

arc A part of a circle whose endpoints are two distinct points of the circle. If the degree measure of the arc is less than 180, the arc is a **minor arc**. If the degree measure of the arc is greater than 180, the arc is a **major arc**. A **semicircle** is an arc whose degree measure is 180.

biconditional A statement of the form "p if and only if q," where statement p is the hypothesis of a conditional statement and statement q is the conclusion. A biconditional is true only when both of its parts have the same truth values.

bisect To divide into two congruent parts.

center of a regular polygon The point in the interior of the polygon that is equidistant from each of the vertices. It is also the common center of its inscribed and circumscribed circles.

central angle of a circle An angle whose vertex is at the center of a circle and whose sides contain radii.

central angle of a regular polygon An angle whose vertex is the center of the polygon and whose sides are drawn to consecutive vertices of the polygon.

centroid of a triangle The point of intersection of its three medians.

chord of a circle A line segment whose endpoints are points on a circle.

circle The set of points (x, y) in the plane that are a fixed distance r from a given point (h, k) called the *center*. An equation of the circle is

$$(x - h)^2 + (y - k)^2 = r^2$$

circumcenter of a triangle The center of the circle that can be circumscribed about a triangle. It can be located by finding the point of intersection of the perpendicular bisectors of two of its sides.

circumscribed circle about a polygon A circle that has each vertex of a polygon on it.

circumscribed polygon about a circle A polygon that has all of its sides tangent to the circle.

collinear points Points that lie on the same line.

common external tangent A line tangent to two circles that does *not* intersect the line joining their centers at a point between the two circles.

common internal tangent A line tangent to two circles that intersects the line joining their centers at a point between the two circles.

complementary angles Two angles whose degree measures add up to 90.

composition of transformations A sequence of transformations in which one transformation is applied to the image of another transformation.

concave polygon A polygon in which there are two points in the interior of the polygon such that the line through them, when extended, intersects the polygon in more than two points. A concave polygon contains at least one interior angle that measures more than 180.

concentric circles Circles in the same plane with the same center but unequal radii.

conclusion In a conditional statement of the form "If . . . then . . . ," the statement that follows "then."

concurrent When three or more lines intersect at the same point.

conditional statement A statement of the form "If p, then q." A conditional statement is always true except in the single instance in which statement p (the hypothesis) is true and statement q (the conclusion) is false.

cone A solid figure with a circular base and a curved lateral surface that joins the base to a point in a different plane called the *vertex*.

congruent angles (or sides) Angles (or sides) that have the same measure. The symbol "is congruent to" is $\cong$.

congruent circles Circles with congruent radii.

congruent parts Pairs of angles or sides that are equal in measure and, as a result, are congruent.

congruent polygons Two polygons with the same number of sides such that all corresponding angles are congruent and all corresponding sides are congruent.

congruent triangles Two triangles that agree in all of their corresponding parts. Two triangles are congruent if any one of the following conditions is true: the three sides of one triangle are congruent to the corresponding sides of the other triangle (SSS $\cong$ SSS); two sides and the included angle of one triangle are congruent to the corresponding parts of the other triangle (SAS $\cong$ SAS); two angles and the included side of one triangle are congruent to the corresponding parts of the other triangle (ASA $\cong$ ASA); two angles and the side opposite one of these angles are congruent to the corresponding parts of the other triangle (AAS $\cong$ AAS).

conjunction A statement that uses "and" to connect two other statements called *conjuncts*. A conjunction is true only when both conjuncts are true.

contrapositive The statement formed by interchanging and then negating the "If" and "then" parts of a conditional statement. Symbolically, the contrapositive of $p \rightarrow q$ is $\sim q \rightarrow \sim p$.

converse The statement formed by interchanging the "If" and "then" parts of a conditional statement. Symbolically, the converse of $p \rightarrow q$ is $q \rightarrow p$.

convex polygon A polygon for which any line that passes through it, provided the line is not tangent to a side or a corner point, intersects the polygon in exactly two points. Each interior angle of a convex polygon measures less than 180. A nonconvex polygon is called a *concave* polygon.

coplanar A term applied to figures that lie in the same plane.

corollary A theorem that can be easily proved from another theorem.

counterexample A specific example that disproves a statement.

cylinder A solid figure with congruent circular bases that lie in parallel planes connected by a curved lateral surface.

deductive reasoning A valid chain of reasoning that uses general principles and accepted facts to reach a specific conclusion.

degree A unit of angle measurement defined as $\dfrac{1}{360}$ of one complete rotation of a ray about its vertex.

diagonal of a polygon A line segment whose endpoints are two nonconsecutive vertices of the polygon.

diameter A chord of a circle that contains the center of the circle.

dilation A transformation in which a figure is enlarged or reduced in size according to a given scale factor.

direct isometry An isometry that preserves orientation.

disjunction A statement that uses "or" to connect two other statements called *disjuncts*. The symbol for disjunction is ∨. A disjunction is true when at least one of the disjuncts is true.

edge of a solid A segment that is the intersection of two faces of the solid.

equidistant A term that means "same distance."

equilateral triangle A triangle in which the three sides have the same length.

equivalence relation A relation in which the reflexive property ($a = a$), symmetric property ($a = b$ and $b = a$), and transitive property (if $a = b$ and $b = c$, then $a = c$) hold. Congruence, similarity, and parallelism are equivalence relations.

exterior angle of a polygon An angle formed by a side of the polygon and the extension of an adjacent side of the polygon.

externally tangent circles Circles that lie on opposite sides of their common tangent.

extremes The first and last terms of a proportion. In the proportion $\dfrac{a}{b} = \dfrac{c}{d}$, the terms a and d are the extremes.

frustrum The part of a cone that remains after a plane parallel to the base of the cone slices off part of the cone below its vertex.

geometric mean See *mean proportional*.

glide reflection The composite of a line reflection and a translation whose direction is parallel to the reflecting line.

great circle The largest circle that can be drawn on a sphere.

hypotenuse In a right triangle, the side opposite the right angle.

hypothesis In a conditional statement of the form "If . . . then . . . ," the statement that follows "If."

image The result of applying a geometric transformation to an object called the *preimage*.

incenter of a triangle The center of the inscribed circle of a triangle that represents the common point at which its three angle bisectors intersect.

indirect proof A deductive method of reasoning that eliminates all but one possibility for the conclusion.

inductive reasoning The process of reasoning from a few specific cases to a broad generalization. Conclusions obtained through inductive reasoning may or may not be true.

inscribed angle An angle of a circle whose vertex is a point on a circle and whose sides are chords.

inscribed circle of a triangle A circle drawn so that the three sides of the triangle are tangent to the circle.

internally tangent circles Tangent circles that lie on the same side of their common tangent.

inverse of a statement The statement formed by negating the "If" and "then" parts of a conditional statement. Symbolically, the inverse of $p \rightarrow q$ is $\sim p \rightarrow \sim q$.

isometry A transformation that preserves distance. Reflections, translations, and rotations are isometries, so they produce congruent iamges. A dilation is *not* an isometry.

isosceles triangle A triangle in which two sides have the same length.

lateral edge An edge of a prism or pyramid that is not a side of a base.

lateral face A face of a prism or pyramid that is not a base.

leg of a right triangle Either of the two sides of a right triangle that include the right angle.

line of symmetry A line that divides a figure into two congruent reflected parts.

linear pair Two adjacent angles whose exterior sides are opposite rays.

line reflection A transformation in which each point P on one side of the reflecting line is paired with a point P' on the opposite side of it so that the reflecting line is the perpendicular bisector of $\overline{PP'}$. If P is on the reflecting line, then P' coincides with P.

line symmetry When a figure can be reflected in a line so that the image coincides with the original figure.

locus The set of points, and only those points, that satisfy a given condition. The plural of "locus" is "loci."

logically equivalent Statements that always have the same truth values. A statement and its contrapositive are logically equivalent.

major arc An arc whose degree measure is greater than 180 and less than 360.

mapping A pairing of the elements of one set with the elements of another set. A mapping between sets A and B is *one-to-one* if every member of A corresponds to exactly one member of B and every member of B corresponds to exactly one member of A.

mean proportional In the proportion $\dfrac{a}{x} = \dfrac{x}{d}$, x is the mean proportional between a and d where $x = \sqrt{ad}$.

midsegment The line segment whose endpoints are the midpoints of two sides of a triangle.

minor arc An arc whose degree measure is between 0 and 180.

n–gon A polygon with n sides.

negation of a statement The statement with the opposite truth value. Symbolically, the negation of statement p is $\sim p$.

noncollinear points Points that do not all lie on the same line.

orthocenter of a triangle The point of intersection of the three altitudes of a triangle.

parallel lines Coplanar lines that do not intersect.

parallelogram A quadrilateral that has two pairs of parallel sides.

parallel postulate Euclid's controversial assumption that through a point not on a line exactly one line can be drawn parallel to the given line.

perpendicular bisector A line that is perpendicular to a line segment at its midpoint.

perpendicular lines Two lines that intersect to form right angles.

point-slope equation of a line An equation of a line with the form $y - b = m(x - a)$, where m is the slope of the line and (a, b) is a point on the line.

point symmetry A figure has point symmetry if after it is rotated 180° (a half-turn) the image coincides with the original figure.

polygon A closed plane figure bounded by line segments that intersect only at their endpoints.

polyhedron A closed solid figure in which each side is a polygon.

postulate A statement that is accepted as true without proof.

preimage The original figure in a transformation. If A' is the image of A under a transformation, then A is the preimage of A'.

prism A polyhedron whose faces, called *bases*, are congruent polygons in parallel planes.

proportion An equation that states that two ratios are equal. In the proportion $\dfrac{a}{b} = \dfrac{c}{d}$, a and d are called the *extremes* and b and c are called the *means*. In a proportion, the product of the means is equal to the product of the extremes. Thus, $a \times d = b \times c$.

pyramid A solid figure formed by joining the vertices of a polygon base to a point in a different plane known as the *vertex*.

Pythagorean Theorem In a right triangle, the square of the length of the hypotenuse is equal to the sum of the squares of the lengths of the two legs.

quadrilateral A polygon with four sides.

ray Part of a line consisting of an endpoint and the set of all points on one side of the endpoint.

rectangle A parallelogram with four right angles.

reflection A transformation that produces an image that is the mirror image of the original object.

reflection rules Reflections of points in the coordinate axes are given by the following rules: $r_{x\text{-axis}}(x, y) = (x, -y)$, $r_{y\text{-axis}}(x, y) = (-x, y)$. To reflect a point in the origin, use the rule $r_{\text{origin}}(x, y) = (-x, -y)$. See also *line reflection*.

reflexive property A quantity is equal (or congruent) to itself.

regular polygon A polygon that is both equilateral and equiangular, such as a square.

rhombus A parallelogram with four congruent sides.

right angle An angle whose degree measure is 90.

right triangle A triangle that contains a right angle.

rotation A transformation in which a point or figure is turned a given number of degrees about a fixed point.

rotation rules The images of points rotated about the origin through angles that are multiples of 90° are given by the following rules: $R_{90°}(x, y) = (-y, x)$, $R_{180°}(x, y) = (-x,-y)$, and $R_{270°}(x, y) = (y, -x)$.

rotational symmetry When a figure can be rotated through a positive angle of less than 360° so that the image coincides with the original figure.

scalene triangle A triangle in which no two sides are congruent.

secant line A line that intersects a circle in two different points.

sector of a circle The interior region of a circle bounded by two radii and their intercepted arc.

segment of a circle The interior region of a circle bounded by a chord of a circle and its intercepted arc.

semicircle An arc whose degree measure is 180.

similar polygons Polygons with the same shape. Similar polygons have congruent corresponding angles and corresponding sides that are in proportion.

similar triangles Triangles with congruent corresponding angles and corresponding sides whose lengths are in proportion. Two triangles are similar if any one of the following conditions is true: (1) two angles of one triangle are congruent to the corresponding angles of the other triangle; (2) a pair of corresponding angles are congruent and the lengths of the sides including those angles are in proportion; (3) the lengths of the corresponding sides of the two triangles are in proportion.

skew lines Lines in different planes that do not intersect but are not parallel.

slant height of a right cone The distance from any point along the circular base to the vertex.

slant height of a pyramid The distance measured along a face from the base to the vertex.

slope A numerical measure of the steepness of a line. The slope of a vertical line is undefined.

slope-intercept equation of a line An equation of a line with the form $y = mx + b$, where m is the slope of the line and b is the y-intercept.

sphere The set of all points in space that are at a fixed distance from a given point called the *center* of the sphere.

square A parallelogram with four right angles and four congruent sides.

substitution property A quantity may be replaced by its equal in an equation.

supplementary angles Two angles whose degree measures add up to 180.

symmetric property If $a = b$, then $b = a$.

tangent circles Circles in the same plane that are tangent to the same line at the same point. *Internally tangent circles* lie on the same side of the common tangent. *Externally tangent circles* lie on opposite sides of the common tangent.

tangent of a circle A line in the same plane as the circle that intersects it in exactly one point.

theorem A mathematical generalization that can be proved.

transformation A change in the position, size, or shape of a figure according to some given rule.

transitive property If $a = b$ and $b = c$, then $a = c$.

translation A transformation in which each point of a figure is shifted the same distance and in the same direction. The notation $T_{h,k}(x, y)$ represents the translation of a point in the coordinate plane h units horizontally and k units vertically. Thus, $T_{h,k}(x, y) = (x + h, y + k)$.

transversal A line that intersects two or more other lines in different points.

trapezoid A quadrilateral in which exactly one pair of sides is parallel. The nonparallel sides are called *legs*.

triangle inequality In any triangle, the length of each side must be less than the sum of the lengths of the other two sides and greater than their difference.

truth value For a statement, either true or false, but not both.

vertex angle In an isosceles triangle, the angle formed by the two congruent sides.

vertex of a polygon The point at which two sides of the polygon intersect. The plural of "vertex" is "vertices."

vertical angles Opposite pairs of congruent angles formed when two lines intersect.

volume The amount of space a solid figure occupies as measured by the number of nonoverlapping $1 \times 1 \times 1$ unit cubes that can exactly fill its interior.

Regents Examinations, Answers, and Self-Analysis Charts

Examination
August 2009
Geometry

PART I

**Answer all 28 questions in this part. Each correct answer will
receive 2 credits. No partial credit will be allowed. For each
question, write in the space provided the number preceding
the word or expression that best completes the statement or
answers the question.** [56 credits]

1 Based on the diagram below, which statement is
true?

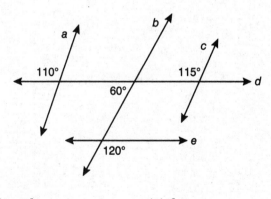

(1) $a \parallel b$ (3) $b \parallel c$

(2) $a \parallel c$ (4) $d \parallel e$

1 __4__

2 The diagram below shows the construction of the bisector of $\angle ABC$.

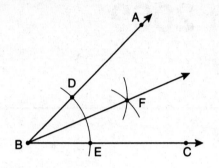

Which statement is *not* true?

(1) $\text{m}\angle EBF = \dfrac{1}{2}\text{m}\angle ABC$

(2) $\text{m}\angle DBF = \dfrac{1}{2}\text{m}\angle ABC$

(3) $\text{m}\angle EBF = \text{m}\angle ABC$

(4) $\text{m}\angle DBF = \text{m}\angle EBF$

2 _3_

3 In the diagram of △ABC below, $\overline{AB} \cong \overline{AC}$. The measure of ∠B is 40°.

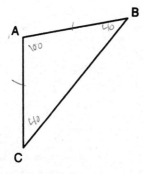

What is the measure of ∠A?

(1) 40°   (3) 70°
(2) 50°   (4) 100°   3 _4_

4 In the diagram of circle O below, chord $\overline{CD}$ is parallel to diameter $\overline{AOB}$ and $m\widehat{AC} = 30$.

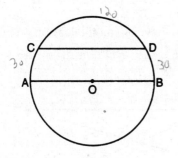

What is $m\widehat{CD}$?

(1) 150   (3) 100
(2) 120   (4) 60   4 _2_

5 In the diagram of trapezoid *ABCD* below, diagonals
$\overline{AC}$ and $\overline{BD}$ intersect at *E* and $\triangle ABC \cong \triangle DCB$.

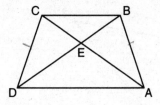

Which statement is true based on the given
information?

(1) $\overline{AC} \cong \overline{BC}$ (3) $\angle CDE \cong \angle BAD$

(2) $\overline{CD} \cong \overline{AD}$ (4) $\angle CDB \cong \angle BAC$ 5 __4__

6 Which transformation produces a figure similar but
not congruent to the original figure?

(1) $T_{1,3}$ (3) $R_{90°}$

(2) $D_{\frac{1}{2}}$ (4) $r_{y=x}$ 6 __2__

7 In the diagram below of parallelogram *ABCD*
with diagonals $\overline{AC}$ and $\overline{BD}$, $m\angle 1 = 45$ and
$m\angle DCB = 120$.

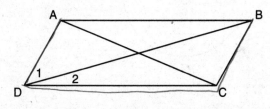

What is the measure of $\angle 2$?

(1) $15°$ (3) $45°$

(2) $30°$ (4) $60°$ 7 __1__

8 On the set of axes below, Geoff drew rectangle *ABCD*. He will transform the rectangle by using the translation $(x,y) \rightarrow (x + 2, y + 1)$ and then will reflect the translated rectangle over the *x*-axis.

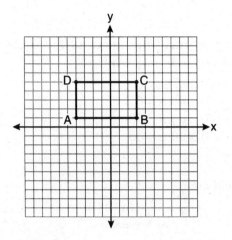

What will be the area of the rectangle after these transformations?

(1) exactly 28 square units
(2) less than 28 square units
(3) greater than 28 square units
(4) It cannot be determined from the information given.

8 ___1___

9 What is the equation of a line that is parallel to the line whose equation is $y = x + 2$?

(1) $x + y = 5$ (3) $y - x = -1$
(2) $2x + y = -2$ (4) $y - 2x = 3$

9 ___3___

10 The endpoints of $\overline{CD}$ are $C(-2,-4)$ and $D(6,2)$. What are the coordinates of the midpoints of $\overline{CD}$?

(1) (2,3)   (3) (4,−2)

(2) (2,−1)   (4) (4,3)   10 __4__

11 What are the center and the radius of the circle whose equation is $(x-3)^2 + (y+3)^2 = 36$?

(1) center = (3,−3); radius = 6

(2) center = (−3,3); radius = 6

(3) center = (3,−3); radius = 36

(4) center = (−3,3); radius = 36   11 __1__

12 Given the equations:

$$y = x^2 - 6x + 10$$
$$y + x = 4$$

What is the solution to the given system of equations?

(1) (2,3)   (3) (2,2) and (1,3)

(2) (3,2)   (4) (2,2) and (3,1)   12 __4__

13 The diagonal $\overline{AC}$ is drawn in parallelogram $ABCD$. Which method can *not* be used to prove that $\triangle ABC \cong \triangle CDA$?

(1) SSS   (3) SSA

(2) SAS   (4) ASA   13 __3__

4 In the diagram below, line k is perpendicular to plane $\mathcal{P}$ at point T.

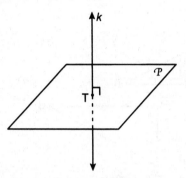

Which statement is true?

(1) Any point in plane $\mathcal{P}$ also will be on line k.
(2) Only one line in plane $\mathcal{P}$ will intersect line k.
(3) All planes that intersect plane $\mathcal{P}$ will pass through T.
(4) Any plane containing line k is perpendicular to plane $\mathcal{P}$.

14 ___4___

15 In the diagram below, which transformation was used to map △*ABC* to △*A'B'C'*?

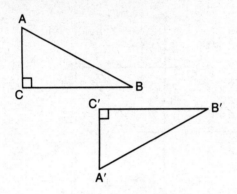

(1) dilation   (3) reflection

(2) rotation   (4) glide reflection   15 __4__

16 Which set of numbers represents the lengths of the sides of a triangle?

  (1) {5, 18, 13}   (3) {16, 24, 7}

  (2) {6, 17, 22}   (4) {26, 8, 15}   16 __2__

17 What is the slope of a line perpendicular to the line whose equation is $y = -\dfrac{2}{3}x - 5$?

  (1) $-\dfrac{3}{2}$ (3) $\dfrac{2}{3}$

  (2) $-\dfrac{2}{3}$ (4) $\dfrac{3}{2}$ 17 __4__

18 A quadrilateral whose diagonals bisect each other and are perpendicular is a

(1) rhombus   (3) trapezoid
(2) rectangle   (4) parallelogram   18 __1__

19 If the endpoints of $\overline{AB}$ are $A(-4,5)$ and $B(2,-5)$, what is the length of $\overline{AB}$?

(1) $2\sqrt{34}$ (3) $\sqrt{61}$
(2) 2   (4) 8   19 __1__

20 In the diagram below of $\triangle ACT$, D is the midpoint of $\overline{AC}$, O is the midpoint of $\overline{AT}$, and G is the midpoint of $\overline{CT}$.

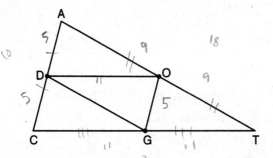

If $AC = 10$, $AT = 18$, and $CT = 22$, what is the perimeter of parallelogram $CDOG$?

(1) 21   (3) 32
(2) 25   (4) 40   20 __3__

21 Which equation represents circle K shown in the graph below?

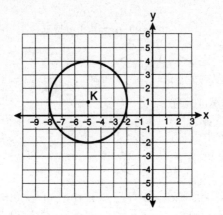

(1) $(x + 5)^2 + (y - 1)^2 = 3$

(2) $(x + 5)^2 + (y - 1)^2 = 9$

(3) $(x - 5)^2 + (y + 1)^2 = 3$

(4) $(x - 5)^2 + (y + 1)^2 = 9$

$\times$ 21 ___3___

22 In the diagram below of right triangle ACB, altitude $\overline{CD}$ is drawn to hypotenuse $\overline{AB}$.

$$\frac{x}{12} = \frac{12}{36}$$

$$\frac{144}{36} = \frac{36x}{36}$$

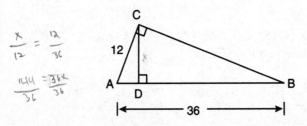

If $AB = 36$ and $AC = 12$, what is the length of $\overline{AD}$?

(1) 32   (3) 3

(2) 6   (4) 4

22 ___4___

23 In the diagram of circle O below, chord $\overline{AB}$ intersects chord $\overline{CD}$ at E, $DE = 2x + 8$, $EC = 3$, $AE = 4x - 3$, and $EB = 4$.

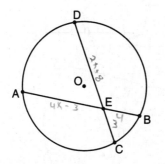

$3(2x+8) = 4(4x-3)$

$6x + 24 = 16x - 12$

$16x - 12 = 6x + 24$

$-6x \qquad -6x$

$10x - 12 = 24$

$\quad +12 \quad +12$

$10x = 36$

$\dfrac{10x}{10} = \dfrac{36}{10}$

What is the value of x?

(1) 1   (3) 5

(2) 3.6   (4) 10.25   23 __2__

24 What is the negation of the statement "Squares are parallelograms"?

(1) Parallelograms are squares.

(2) Parallelograms are not squares.

(3) It is not the case that squares are parallelograms.

(4) It is not the case that parallelograms are squares.   24 __3__

25 The diagram below shows the construction of the center of the circle circumscribed about $\triangle ABC$.

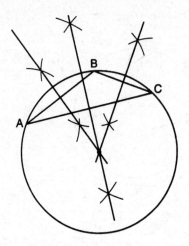

This construction represents how to find the intersection of

(1) the angle bisectors of $\triangle ABC$
(2) the medians to the sides of $\triangle ABC$
(3) the altitudes to the sides of $\triangle ABC$
(4) the perpendicular bisectors of the sides of $\triangle ABC$

25 __4__

26 A right circular cylinder has a volume of 1,000 cubic inches and a height of 8 inches. What is the radius of the cylinder to the *nearest tenth of an inch*?

(1) 6.3   (3) 19.8
(2) 11.2   (4) 39.8

26 __1__

27 If two different lines are perpendicular to the same plane, they are

(1) collinear   (3) congruent
(2) coplanar   (4) consecutive

27 __2__

28 How many common tangent lines can be drawn to the two externally tangent circles shown below?

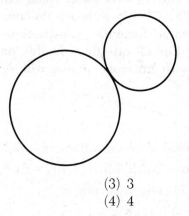

(1) 1

(2) 2

(3) 3

(4) 4

28 _3_

PART II

Answer all 6 questions in this part. Each correct answer will receive 2 credits. Clearly indicate the necessary steps, including appropriate formula substitutions, diagrams, graphs, charts, etc. For all questions in this part, a correct numerical answer with no work shown will receive only 1 credit.   [12 credits]

29 In the diagram below of isosceles trapezoid $DEFG$, $\overline{DE} \parallel \overline{GF}$, $DE = 4x - 2$, $EF = 3x + 2$, $FG = 5x - 3$, and $GD = 2x + 5$. Find the value of x.

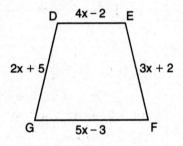

$$2x + 5 = 3x + 2$$
$$3x + 2 = 2x + 5$$
$$-2x \qquad -2x$$
$$x + 2 = 5$$
$$-2 \quad -2$$
$$\boxed{x = 3}$$

30 A regular pyramid with a square base is shown in the diagram below.

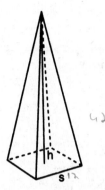

42

$\dfrac{144\,(42)}{3}$ = 2016 M³

h

s 12

A side, *s*, of the base of the pyramid is 12 meters, and the height, *h*, is 42 meters. What is the volume of the pyramid in cubic meters?

31 Write an equation of the line that passes through the point (6,–5) and is parallel to the line whose equation is $2x - 3y = 11$.

$2x - 3y = 11$

$\dfrac{-3y}{-3} = \dfrac{-2 \cdot x}{-3} + \dfrac{11}{-3}$

$y = \dfrac{2}{3}x + b$

$\boxed{y = \dfrac{2}{3}x - 9}$

$-5 = \dfrac{2}{3}(6) + b$

$-5 = 4 + b$

$-4 \quad -4$

$-9 = b$

$b = -9$

32 Using a compass and straightedge, construct the angle bisector of ∠ABC shown below. [Leave all construction marks.]

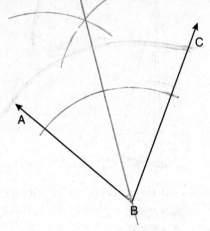

33 The degree measures of the angles of △ABC are represented by x, $3x$, and $5x - 54$. Find the value of x.

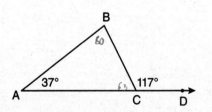

$\boxed{x = 26}$ $x + 3x + 5x - 54 = 180$ $9x = 234$

$9x \overset{+54}{=} \overset{+54}{180}$ $\frac{9x}{9} = \frac{234}{9}$

$9x = 234$

34 In the diagram below of △ABC with side $\overline{AC}$ extended through D, m∠A = 37 and m∠BCD = 117. Which side of △ABC is the longest side? Justify your answer.

$117 - 37 = 80$

$180 - 117 = 63$

AC is largest

Since opposite of largest angle

B

60

A   37°   63   117°   C   D

(Not drawn to scale)

PART III

Answer all 3 questions in this part. Each correct answer will receive 4 credits. Clearly indicate the necessary steps, including appropriate formula substitutions, diagrams, graphs, charts, etc. For all questions in this part, a correct numerical answer with no work shown will receive only 1 credit. [12 credits]

$$\frac{x_1 + x_2}{2} \quad , \quad \frac{-1, 1 + 2}{2} \quad \frac{7-1}{2}, \quad \frac{-5+1}{2}$$

$$\frac{6}{2} \qquad \frac{-4}{2}$$

35 Write an equation of the perpendicular bisector of the line segment whose endpoints are $(-1,1)$ and $(7,-5)$. [The use of the grid below is optional.]

$$3, -2$$

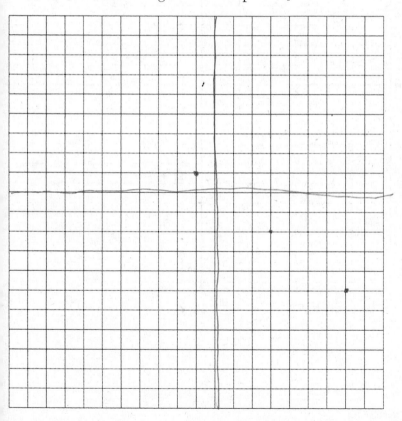

36 On the set of axes below, sketch the points that are 5 units from the origin and sketch the points that are 2 units from the line $y = 3$. Label with an **X** all points that satisfy *both* conditions.

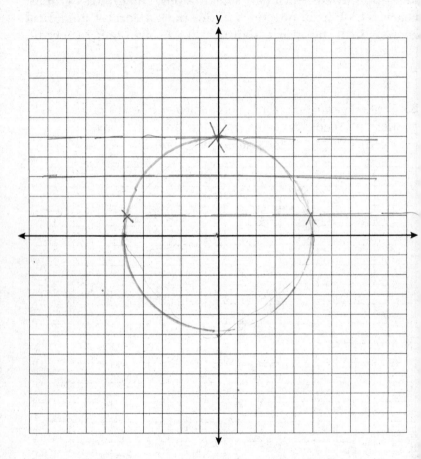

37 Triangle *DEG* has the coordinates $D(1,1)$, $E(5,1)$, and $G(5,4)$. Triangle *DEG* is rotated 90° about the origin to form $\triangle D'E'G'$. On the grid below, graph and label $\triangle DEG$ and $\triangle D'E'G'$. State the coordinates of the vertices D', E', and G'. Justify that this transformation preserves distance.

$D' = -1,1$
$E' = -1,5$
$G' = -4,5$

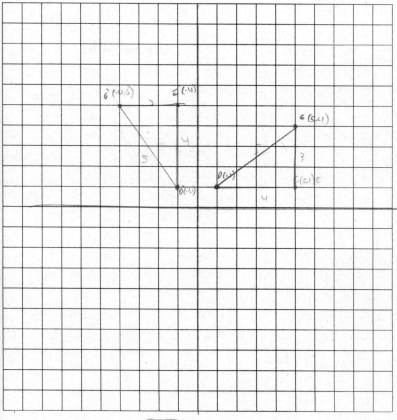

$\overline{GE} = \sqrt{(4-1)^2 + (5-5)^2} = 9$

$3 \quad 9 \to 0$

$\overline{G'E'} = \sqrt{(5-5)^2 + (-1+4)^2}$

$0 + 3^2$

PART IV

Answer the question in this part. A correct answer will receive 6 credits. Clearly indicate the necessary steps, including appropriate formula substitutions, diagrams, graphs, charts, etc. A correct numerical answer with no work shown will receive only 1 credit. [6 credits]

38 Given: Quadrilateral $ABCD$, diagonal $\overline{AFEC}$, $\overline{AE} \cong \overline{FC}$, $\overline{BF} \perp \overline{AC}$, $\overline{DE} \perp \overline{AC}$, $\angle 1 \cong \angle 2$

Prove: $ABCD$ is a parallelogram.

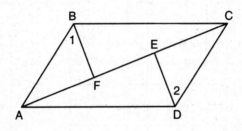

Answers
August 2009
Geometry

Answer Key

PART I

1. (4)	**8.** (1)	**15.** (4)	**22.** (4)
2. (3)	**9.** (3)	**16.** (2)	**23.** (2)
3. (4)	**10.** (2)	**17.** (4)	**24.** (3)
4. (2)	**11.** (1)	**18.** (1)	**25.** (4)
5. (4)	**12.** (4)	**19.** (1)	**26.** (1)
6. (2)	**13.** (3)	**20.** (3)	**27.** (2)
7. (1)	**14.** (4)	**21.** (2)	**28.** (3)

PART II

29. 3

30. 2,016

31. $y + 5 = \frac{2}{3}(x - 6)$ or $y = \frac{2}{3}x - 9$

32. See the *Answers Explained* section.

33. $\frac{26}{}$

34. $\overline{AC}$

PART III

35. $y + 2 = \frac{4}{3}(x - 3)$

   or $y = \frac{4}{3}x - 6$

36. See the *Answers Explained* section.

37. $D'(-1,1)$, $E'(-1,5)$, $G'(-4,5)$

PART IV

38. See the *Answers Explained* section.

In **PARTS II–IV** you are required to show how you arrived at your answers. For sample methods of solutions, see the *Answers Explained* section.

Answers Explained

PART I

   1. In the accompanying diagram, m∠1 = 180 − 60 = 120. Because transversal *b* intersects lines *d* and *e*, the corresponding angles marked in the diagram are congruent, *d* ∥ *e*.

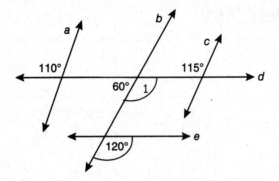

The correct choice is (**4**).

2. It is given that the accompanying diagram represents the construction of the bisector of $\angle ABC$.

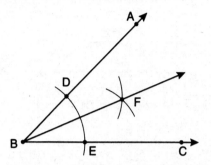

You are asked to choose the statement among the answer choices that is *not* true. Choice (1) states that $m\angle EBF = \frac{1}{2} m\angle ABC$, and choice (3) states that $m\angle EBF = m\angle ABC$. Both statements cannot be true at the same time. Since each angle formed by an angle bisector is one-half of the measure of the original angle, $m\angle EBF = \frac{1}{2} m\angle ABC$ and $m\angle EBF \neq m\angle ABC$.

The correct choice is (**3**).

3. It is given that in the accompanying diagram of $\triangle ABC$, $\overline{AB} \cong \overline{AC}$ and $m\angle B = 40°$.

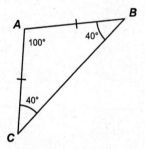

- If two sides of a triangle are congruent, the angles opposite these sides are congruent. Thus $m\angle C = m\angle B = 40°$.
- Since the sum of the measures of the three angles of a triangle is 180°, $m\angle A + 40° + 40° = 180°$ so $m\angle A = 180° - 80°$ and $m\angle A = 100°$.

The correct choice is (**4**).

4. It is given that in the accompanying diagram of circle O, chord $\overline{CD}$ is parallel to diameter $\overline{AOB}$ and $m\widehat{AC} = 30$.

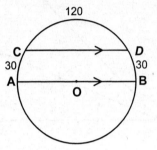

- Parallel chords cut off congruent arcs, so $m\widehat{BD} = m\widehat{AC} = 30$.
- Since a diameter divides a circle into two semicircles, each of which measures 180 degrees:

$$m\widehat{CD} + m\widehat{BD} + m\widehat{AC} = 180$$
$$m\widehat{CD} + 30 + 30 = 180$$
$$m\widehat{CD} = 180 - 60 = 120$$

The correct choice is **(2)**.

5. It is given that in the accompanying diagram of trapezoid *ABCD*, diagonals $\overline{AC}$ and $\overline{BD}$ intersect at *E* and $\triangle ABC \cong \triangle DCB$.

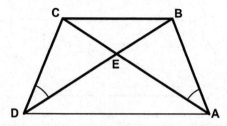

To determine the true statement, find the answer choice that contains the pair of corresponding parts that are congruent as a result of $\triangle ABC \cong \triangle DCB$. Angles *CDB* and *BAC* are corresponding angles in the two overlapping congruent triangles, so $\angle CDB \cong \angle BAC$.

The correct choice is (**4**).

6. The notation $D_{\frac{1}{2}}$ represents a dilation with a scale factor of $\frac{1}{2}$. The distance between any two points of the transformed figure is one-half of the distance between the two corresponding points of the original figure. Since a dilation preserves angle measures, the transformation $D_{\frac{1}{2}}$ produces a figure similar but not congruent to the original figure.

The correct choice is (**2**).

7. It is given that in the accompanying diagram of parallelogram $ABCD$ with diagonals $\overline{AC}$ and $\overline{BD}$, m$\angle 1 = 45$ and m$\angle DCB = 120$.

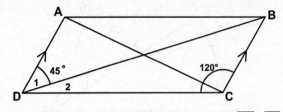

- Since opposite sides of a parallelogram are parallel, $\overline{AD} \parallel \overline{BC}$.
- Since consecutive angles of a parallelogram are supplementary:

$$\text{m}\angle CDA + \text{m}\angle DCB = 180°$$
$$\text{m}\angle 1 + \text{m}\angle 2 + \text{m}\angle DCB = 180°$$
$$45° + \text{m}\angle 2 + 120° = 180°$$
$$\text{m}\angle 2 = 180° - 165°$$
$$= 15°$$

The correct choice is **(1)**.

8. It is given that the translation $(x,y) \rightarrow (x + 2, y + 1)$ will be performed on the rectangle shown on the accompanying set of axes. The translated rectangle will then be reflected over the x-axis. You are asked to find the area of the rectangle after these transformations.

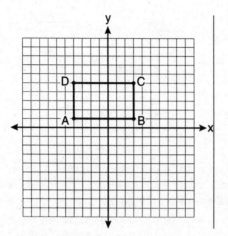

The area of rectangle $ABCD$ is

$$AB \times AD = 7 \times 4 = 28 \text{ square units}$$

Since distance is preserved under both a translation and a reflection, the dimensions of the rectangle after these two transformations will be the same as those of the original rectangle. So the area of the transformed rectangle will also be 28 square units.

The correct choice is **(1)**.

9. You are asked to identify the equation of a line that is parallel to the line whose equation is $y = x + 2$. Since the given line is written in the slope-intercept form of $y = mx + b$, its slope is 1, the coefficient of x. Parallel lines have the same slope. Check each of the answer choices until you find the equation whose slope, m, is 1:

- Choice (1): $x + y = 5$ so $y = -x + 5$ and $m = -1$. ✗
- Choice (2): $2x + y = -2$ so $y = -2x + -2$ and $m = -2$. ✗
- Choice (3): $y - x = -1$ so $y = x - 1$ and $m = 1$. ✓
- Choice (4): $y - 2x = 3$ so $y = 2x + 3$ and $m = 2$. ✗

The correct choice is **(3)**.

10. It is given that the endpoints of $\overline{CD}$ are $C(-2,-4)$ and $D(6,2)$. To find the coordinates of the midpoint $(\bar{x}, \bar{y})$ of $\overline{CD}$, find the averages of the corresponding coordinates of the endpoints:

$$\bar{x} = \frac{-2+6}{2} \qquad\qquad \bar{y} = \frac{-4+2}{2}$$

$$= \frac{4}{2} \qquad \text{and} \qquad = \frac{-2}{2}$$

$$= 2 \qquad\qquad\qquad = -1$$

The coordinates of the midpoint of $\overline{CD}$ are $(2,-1)$.

The correct choice is **(2)**.

11. The center-radius form of an equation of a circle with its center at (h,k) and a radius of r is $(x - h)^2 + (y - k)^2 = r^2$. You are required to identify the center and radius of the circle whose equation is $(x - 3)^2 + (y + 3)^2 = 36$. Rewrite the equation in center-radius form as $(x - 3)^2 + (y - (-3))^2 = 6^2$. Since $(h,k) = (3,-3)$ and $r = 6$, the center of the required circle is at $(3,-3)$ and its radius is 6.

The correct choice is **(1)**.

12. To solve the given system of equations, $y = x^2 - 6x + 10$ and $y + x = 4$, eliminate the variable y in the linear equation using substitution:

$$y = -x + 4$$
$$x^2 - 6x + 10 = -x + 4$$

Write the quadratic equation in standard form so that all of the nonzero terms are on the same side of the equation: $x^2 - 5x + 6 = 0$

Factor the quadratic trinomial as the product of two binomials: $(x - 2)(x - 3) = 0$

Set each factor equal to 0: $x - 2 = 0 \quad \text{or} \quad x - 3 = 0$

$x = 2 \quad \text{or} \quad x = 3$

Find the corresponding values of y by substituting each of the solution values of x into the equation $y + x = 4$:

- When $x = 2$, $y + 2 = 4$ so $y = 2$.
- When $x = 3$, $y + 3 = 4$ so $y = 1$.

The solution to the given system of equations is $(2,2)$ and $(3,1)$.

The correct choice is **(4)**.

13. It is given that diagonal $\overline{AC}$ is drawn in parallelogram $ABCD$. You are asked to identify the method that cannot be used to prove that $\triangle ABC \cong \triangle CDA$. Since SSA is *never* a valid method for proving triangles congruent and this is offered in choice (3), no further analysis is required.

The correct choice is **(3)**.

14. It is given that line k is perpendicular to plane $\mathcal{P}$ at point T.

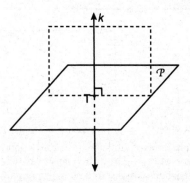

If a line is perpendicular to a plane, then every plane containing the line must also be perpendicular to the given plane.

The correct choice is **(4)**.

15. You are asked to identify the transformation used to map $\triangle ABC$ to $\triangle A'B'C'$.

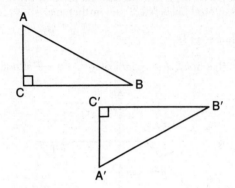

Examine each answer choice.

- Choice (1): A dilation is a similarity transformation that changes the size but not the shape of the figure. The transformation shown is not a dilation as the two triangles appear to be congruent. ✗

- Choice (2): A rotation turns a figure while maintaining orientation. When the vertices of the figure and its rotated image are both read in a clockwise direction, the lettered vertices appear in the same order. In the given transformation, the original figure and its image have reverse orientations. ✗

- Choice (3): A reflection in a line flips a figure over a line while reversing orientation. Since $\triangle A'B'C'$ appears to be shifted in the horizontal direction relative to $\triangle ABC$, the transformation is not a simple reflection. ✗

- Choice (4): A glide reflection is comprised of a reflection and a translation in a direction parallel to the reflecting line. Reflecting $\triangle ABC$ in a line parallel to $\overline{BC}$ and then translating the reflected image horizontally to the right maps $\triangle ABC$ onto $\triangle A'B'C'$. ✓

The correct choice is (**4**).

16. You are asked to identify the set of three numbers that represents the lengths of the sides of a triangle. Examine each answer choice in turn until you find the set of three positive numbers for which each number is less than the sum of the other two, as in choice (2):

$$6 < 17 + 22 \quad \checkmark \quad 17 < 6 + 22 \quad \checkmark \quad 22 < 6 + 17 \quad \checkmark$$

The correct choice is **(2)**.

17. Perpendicular lines have slopes that are negative reciprocals. The slope of the given line, $y = -\dfrac{2}{3}x - 5$, is $-\dfrac{2}{3}$. The negative reciprocal of $-\dfrac{2}{3}$ is $\dfrac{3}{2}$. The slope of a line perpendicular to the given line is, therefore, $\dfrac{3}{2}$.

The correct choice is **(4)**.

18. If the diagonals of a quadrilateral bisect each other, the quadrilateral is a parallelogram. If the diagonals of a parallelogram are perpendicular, the parallelogram is a rhombus.

The correct choice is **(1)**.

19. It is given that the endpoints of $\overline{AB}$ are $A(-4,5)$ and $B(2,-5)$. To find the length of $\overline{AB}$, use the distance formula:

$$
\begin{aligned}
AB &= \sqrt{\left(x_B - x_A\right)^2 + \left(y_B - y_A\right)^2} \\
&= \sqrt{\left(2 - (-4)\right)^2 + \left(-5 - 5\right)^2} \\
&= \sqrt{\left(2 + 4\right)^2 + \left(-10\right)^2} \\
&= \sqrt{36 + 100} \\
&= \sqrt{136} \\
&= \sqrt{4} \cdot \sqrt{34} \\
&= 2\sqrt{34}
\end{aligned}
$$

The correct choice is **(1)**.

20. In the accompanying diagram of $\triangle ACT$, it is given that D is the midpoint of $\overline{AC}$, O is the midpoint of $\overline{AT}$, and G is the midpoint of $\overline{CT}$. It is also given that $AC = 10$, $AT = 18$, and $CT = 22$. You are required to find the perimeter of parallelogram $CDOG$.

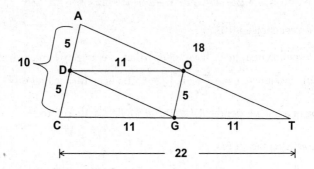

- Because D and G are midpoints, $CD = \dfrac{1}{2}AC = \dfrac{1}{2}(10) = 5$ and $CG = \dfrac{1}{2}CT = \dfrac{1}{2}(22) = 11.$

- Since opposite sides of a parallelogram are congruent, $GO = CD = 5$ and $OD = CG = 11.$

- To find the perimeter of $CDOG$, find the sum of the lengths of its four sides: $5 + 11 + 5 + 11 = 32.$

The correct choice is **(3)**.

21. You are asked to identify the equation of circle K in the accompanying graph.

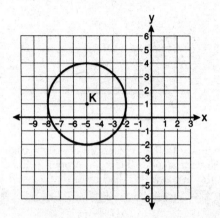

Reading from the graph, the radius $r = 3$ and the center (h,k) is at $(-5,1)$. Substitute these values into the center-radius form of the equation of a circle:

$$(x - h)^2 + (y - k)^2 = r^2$$
$$(x - (-5))^2 + (y - 1)^2 = 3^2$$
$$(x + 5)^2 + (y - 1)^2 = 9$$

The correct choice is **(2)**.

22. In the accompanying diagram of right triangle ACB, altitude $\overline{CD}$ is drawn to hypotenuse $\overline{AB}$. You must find the length of $\overline{AD}$ given that $AB = 36$ and $AC = 12$.

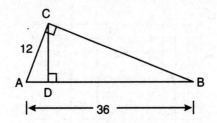

If an altitude is drawn to the hypotenuse of a right triangle, the length of each leg of the original triangle is the mean proportional between the length of the segment of the hypotenuse adjacent to that leg and the length of the whole hypotenuse:

$$\frac{AD}{AC} = \frac{AC}{AB}$$

$$\frac{AD}{12} = \frac{12}{36}$$

$$36\,(AD) = 12 \cdot 12$$

$$\frac{36(AD)}{36} = \frac{144}{36}$$

$$AD = 4$$

The correct choice is **(4)**.

23. In the accompanying diagram of circle O, chord $\overline{AB}$ intersects chord $\overline{CD}$ at E, $DE = 2x + 8$, $EC = 3$, $AE = 4x - 3$, and $EB = 4$.

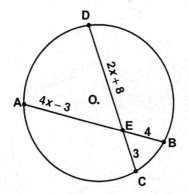

If two chords intersect inside a circle, then the product of the segment lengths of one chord is equal to the product of the segment lengths of the other chord:

$$EB \times AE = EC \times DE$$
$$4(4x - 3) = 3(2x + 8)$$
$$16x - 12 = 6x + 24$$
$$16x - 6x = 24 + 12$$
$$10x = 36$$
$$\frac{10x}{10} = \frac{36}{10}$$
$$x = 3.6$$

The correct choice is (**2**).

24. To find the negation of the statement "Squares are parallelograms," form the statement that has the opposite truth value, "It is not the case that squares are parallelograms."

The correct choice is (**3**).

25. The accompanying diagram shows the construction of the center of the circle circumscribed about △*ABC*.

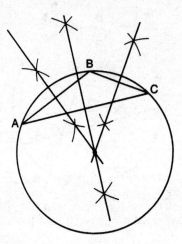

The perpendicular bisectors of the sides of a triangle meet at a point that is equidistant from the vertices of the triangle, which is the center of the circle circumscribed about △*ABC*.

The correct choice is (**4**).

26. It is given that a right circular cylinder has a volume of 1,000 cubic inches and a height of 8 inches. To find the radius, *r*, of the cylinder, use your calculator and the volume formula $V = \pi r^2 h$, where $V = 1{,}000$ and $h = 8$:

$$1{,}000 = \pi r^2(8)$$

$$r^2 = \frac{1{,}000}{8\pi}$$

$$r = \sqrt{\frac{1{,}000}{8\pi}}$$

$$r \approx \sqrt{39.78873577}$$

$$r \approx 6.307831305$$

To the *nearest tenth of an inch*, $r = 6.3$.

The correct choice is (**1**).

27. In the accompanying diagram, line $m \perp$ plane P and line $n \perp$ plane P, so lines m and n must both lie in plane Q. In general, if two different lines are perpendicular to the same plane, the two lines must lie in the same plane. As a result, the lines are coplanar.

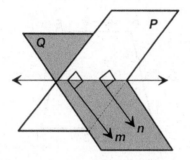

The correct choice is **(2)**.

28. You are asked to determine the number of common tangent lines that can be drawn to two externally tangent circles. If two circles are externally tangent, then two common external tangents and one common internal tangent can be drawn, as shown in the accompanying diagram:

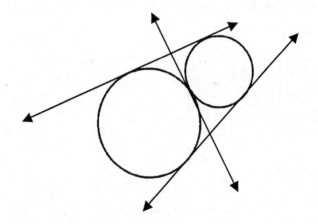

Thus, a total of three common tangent lines can be drawn.

The correct choice is **(3)**.

PART II

29. In the diagram below of isosceles trapezoid $DEFG$, $\overline{DE} \parallel \overline{GF}$, $DE = 4x - 2$, $EF = 3x + 2$, $FG = 5x - 3$, and $GD = 2x + 5$.

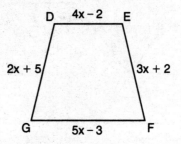

Since the nonparallel sides of an isosceles trapezoid are congruent,

$$EF = GD$$
$$3x + 2 = 2x + 5$$
$$3x - 2x = 5 - 2$$
$$x = 3$$

The value of x is **3**.

30. A regular pyramid with a square base is shown in the diagram below. A side, s, of the base of the pyramid is 12 meters. The height, h, is 42 meters. You are required to find the volume of the pyramid.

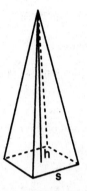

The volume, V, of a pyramid is one-third of the area of the base, denoted by B, times the height, h. Since the base of the pyramid is a square, its area is $s \times s$:

$$V = \frac{1}{3}Bh$$
$$= \frac{1}{3}(s \times s)h$$
$$= \frac{1}{3}(12\,\text{m} \times 12\,\text{m})\,42\,\text{m}$$
$$= 48 \times 42\,\text{m}^3$$
$$= 2{,}016\,\text{m}^3$$

The volume of the pyramid is **2,016** cubic meters.

31. Write an equation of the line that passes through the point $(6,-5)$ and is parallel to the line whose equation is $2x - 3y = 11$.

- Find the slope of the given line by writing its equation in slope-intercept form. If $2x - 3y = 11$, then $-3y = -2x + 11$ so $y = \dfrac{2}{3}x - \dfrac{11}{3}$. The slope of the line is $\dfrac{2}{3}$, the coefficient of x.

- Since parallel lines have the same slope, the slope, m, of the required line is $\dfrac{2}{3}$.

<u>Method 1</u>: Write an equation of the required line using the point-slope form $y - y_1 = m(x - x_1)$ where $(x_1, y_1) = (6,-5)$ and $m = \dfrac{2}{3}$:

$$y - (-5) = \frac{2}{3}(x - 6)$$

$$\boldsymbol{y + 5 = \frac{2}{3}(x - 6)}$$

<u>Method 2</u>: Write an equation of the required line using the slope-intercept form where $m = \dfrac{2}{3}$ and b is determined by substituting the coordinates $(6,-5)$ into the equation $y = mx + b$:

$$-5 = \frac{2}{\cancel{3}}(\cancel{6}^{2}) + b$$

$$-5 = 4 + b$$

$$-9 = b$$

Because $m = \dfrac{2}{3}$ and $b = -9$, an equation of the required line is $\boldsymbol{y = \dfrac{2}{3}x - 9}$.

32. Using a compass and straightedge, construct the bisector of ∠*ABC* shown below.

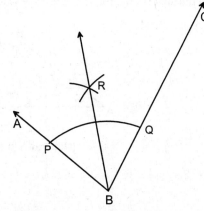

- Step 1: Using *B* as a center and any convenient radius length, construct an arc that intersects the sides of the angle. Label these points *P* and *Q*.

- Step 2: Using points *P* and *Q* as centers, construct a pair of arcs that intersect. Label the point of intersection *R*.

- Step 3: Draw $\overrightarrow{BR}$.

Hence, $\overrightarrow{BR}$ is the bisector of ∠*ABC*.

33. The degree measures of the angles of △*ABC* are represented by x, $3x$, and $5x - 54$. To find the value of x, add the measures of the three angles together and set the sum equal to 180:

$$x + 3x + \left(5x - 54\right) = 180$$
$$9x - 54 = 180$$
$$9x = 180 + 54$$
$$\frac{9x}{9} = \frac{234}{9}$$
$$x = 26$$

The value of x is **26**.

34. In the accompanying diagram of $\triangle ABC$, side $\overline{AC}$ is extended through D, m$\angle A = 37$, and m$\angle BCD = 117$. To determine the longest side of $\triangle ABC$, find the angle of the triangle with the greatest measure.

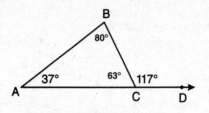

(Not drawn to scale)

- Because $\angle ACB$ and $\angle BCD$ are supplementary:

$$m\angle ACB = 180 - m\angle BCD$$
$$= 180 - 117$$
$$= 63$$

- Since the sum of the measures of the angles of a triangle is 180:

$$37 + m\angle B + 63 = 180$$
$$m\angle B = 180 - 100$$
$$= 80$$

- In a triangle, the longest side is opposite the angle with the greatest measure. In $\triangle ABC$, angle B has the greatest measure. Since $\overline{AC}$ is opposite $\angle B$, $\overline{AC}$ is the longest side of the triangle.

The longest side of the triangle is $\overline{AC}$.

PART III

35. You are required to write an equation of the perpendicular bisector of the line segment whose endpoints are $(-1,1)$ and $(7,-5)$.

- The perpendicular bisector passes through the midpoint of the line segment whose endpoints are $(-1,1)$ and $(7,-5)$. Find the coordinates of the midpoint $M(\bar{x},\bar{y})$ by taking the averages of the corresponding coordinates of the endpoints:

$$\bar{x} = \frac{-1+7}{2} \qquad\qquad \bar{y} = \frac{1-5}{2}$$

$$= \frac{6}{2} \qquad \text{and} \qquad = \frac{-4}{2}$$

$$= 3 \qquad\qquad\qquad = -2$$

The coordinates of the midpoint are $M(3,-2)$.

- Use the slope formula to find the slope of the line whose endpoints are $(-1,1)$ and $(7,-5)$:

$$\text{slope} = \frac{\Delta y}{\Delta x}$$

$$= \frac{-5-1}{7-(-1)}$$

$$= \frac{-6}{8}$$

$$= -\frac{3}{4}$$

Since the slopes of perpendicular lines are negative reciprocals, the slope of the perpendicular bisector is $\frac{4}{3}$.

<u>Method 1</u>: Write an equation of the perpendicular bisector using the point-slope form $y - y_1 = m(x - x_1)$ where $(x_1, y_1) = M(3,-2)$ and $m = \frac{4}{3}$:

$$y - (-2) = \frac{4}{3}(x - 3)$$

$$y + 2 = \frac{4}{3}(x - 3)$$

<u>Method 2</u>: Write an equation of the required line using the slope-intercept form where $m = \dfrac{4}{3}$ and b is determined by substituting the coordinates $M(3,-2)$ into the equation $y = mx + b$:

$$-2 = \frac{4}{\cancel{3}}\left(\cancel{3}^{1}\right) + b$$
$$-2 = 4 + b$$
$$-6 = b$$

Because $m = \dfrac{4}{3}$ and $b = -6$, an equation of the line is $\boldsymbol{y = \dfrac{4}{3}x - 6}$.

36. Using the accompanying set of axes, you are asked to sketch the points that are 5 units from the origin, sketch the points that are 2 units from the line $y = 3$, and label with an **X** all points that satisfy *both* conditions.

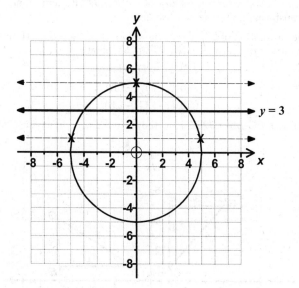

- To sketch the set of all points that are 5 units from the origin, draw a circle whose center is at the origin and has a radius of 5 units, as shown on the accompanying set of axes.

- To sketch the set of all points that are 2 units from the line $y = 3$, draw parallel lines on either side of that line that are 2 units from it. In other words, draw a solid horizontal line through $y = 3$ and then draw broken horizontal lines through $y = 1$ and $y = 5$, as shown on the accompanying set of axes.

- The line $y = 1$ intersects the circle at two points. Since the radius of the circle is 5, the line $y = 5$ intersects the circle at exactly one point. Mark each of the three points with an **X**, as shown in the diagram. These three points satisfy *both* of the given conditions.

37. Triangle *DEC* has the coordinates *D*(1,1), *E*(5,1), and *G*(5,4). Triangle *DEC* is rotated 90° about the origin to form △*D′E′G′*. On the accompanying set of axes, you are asked to graph and label △*DEG* and △*D′E′G′*. You are also required to state the coordinates of vertices *D′*, *E′*, and *G′* and to justify that this transformation preserves distance.

- Under a rotation of 90° about the origin, (*x,y*) → (−*y,x*). Hence, the co-ordinates of the vertices △*D′E′G′* are as follows:

$$D(1,1) \rightarrow \textbf{\textit{D}}'\textbf{(--1,1)}$$
$$E(5,1) \rightarrow \textbf{\textit{E}}'\textbf{(--1,5)}$$
$$G(5,4) \rightarrow \textbf{\textit{G}}'\textbf{(--4,5)}$$

- Graph △*DEG* and △*D′E′G′*, as shown in the accompanying figure.

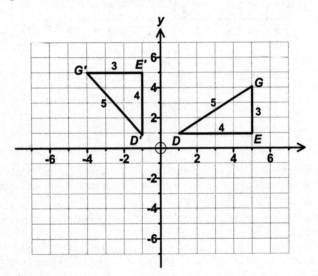

- By counting grid boxes, you know that *GE* = *G′E′* = 3 and that *ED* = *E′D′* = 4. Since △*DEG* and △*D′E′G′* are 3-4-5 right triangles, *DG* = *D′G′* = 5. Since corresponding sides of the two triangles have the same length, this transformation preserves distance.

PART IV

38. In the accompanying diagram, you are given quadrilateral *ABCD*, diagonal $\overline{AFEC}$, $\overline{AE} \cong \overline{FC}$, $\overline{BF} \perp \overline{AC}$, $\overline{DE} \perp \overline{AC}$, and $\angle 1 \cong \angle 2$. You are required to prove that *ABCD* is a parallelogram.

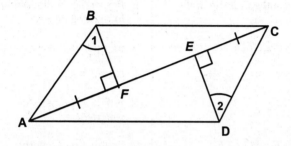

Plan: A quadrilateral is a parallelogram if one pair of sides are both congruent and parallel. Prove $\triangle AFB \cong \triangle CED$. Then use CPCTC to show that $\overline{AB} \cong \overline{CD}$ and $\overline{AB} \parallel \overline{CD}$.

Statement	Reason
1. Quadrilateral *ABCD*, diagonal $\overline{AFEC}$, $\overline{BF} \perp \overline{AC}$, $\overline{DE} \perp \overline{AC}$, $\angle 1 \cong \angle 2$. *Angle*	1. Given.
2. $\angle AFB$ and $\angle CED$ are right angles.	2. Perpendicular lines intersect to form right angles.
3. $\angle AFB \cong \angle CED$. *Angle*	3. All right angles are congruent.
4. $\overline{AE} \cong \overline{FC}$.	4. Given.
5. $\overline{EF} \cong \overline{EF}$.	5. Reflexive property of congruence.
6. $AE - EF = FC - EF$.	6. Subtraction property.
7. $\overline{AF} \cong \overline{CE}$. *Side*	7. Substitution.
8. $\triangle AFB \cong \triangle CED$.	8. AAS.
9. $\overline{AB} \cong \overline{CD}$.	9. Corresponding parts of congruent triangles are congruent (CPCTC).
10. $\angle FAB \cong \angle ECD$.	10. Corresponding parts of congruent triangles are congruent.
11. $\overline{AB} \parallel \overline{CD}$.	11. If two lines intersect to form congruent alternate interior angles, the lines are parallel.
12. *ABCD* is a parallelogram.	12. If a quadrilateral has one pair of sides that are congruent and parallel, it is a parallelogram.

Topic	Question Numbers	Number of Points	Your Points	Your Percentage
1. Logic (negation, conjunction, disjunction, related conditionals, biconditional, logically equivalent statements, logical inference)	24	2		
2. Angle & Line Relationships (vertical angles, midpoint, altitude, median, bisector, supplementary angles, complementary angles, parallel and perpendicular lines)	1, 2	2 + 2 = 4		
3. Parallel & Perpendicular Planes	14, 27	2 + 2 = 4		
4. Angles of a Triangle & Polygon (sum of angles, exterior angle, base angles theorem, angles of a regular polygon)	3, 33	2 + 2 = 4		
5. Triangle Inequalities (side length restrictions, unequal angles opposite unequal sides, exterior angle)	16, 34	2 + 2 = 4		
6. Trapezoids & Parallelograms (includes properties of rectangle, rhombus, square, and isosceles trapezoid)	5, 7, 18, 29	2 + 2 + 2 + 2 = 8		
7. Proofs Involving Congruent Triangles	13, 38	2 + 6 = 8		
8. Indirect Proof & Mathematical Reasoning	—	—		
9. Ratio & Proportion (includes similar polygons, proportions, proofs involving similar triangles, comparing linear dimensions, and areas of similar triangles)	—	—		
10. Proportions formed by altitude to hypotenuse of right triangle; Pythagorean theorem	22	2		
11. Coordinate Geometry (slopes of parallel and perpendicular lines; slope-intercept and point-slope equations of a line; applications requiring midpoint, distance, or slope formulas; coordinate proofs)	9, 10, 17, 19, 31, 35	2 + 2 + 2 + 2 + 2 + 4 = 14		
12. Solving a Linear-Quadratic System of Equations Graphically	12	2		
13. Transformation Geometry (includes isometries, dilation, symmetry, composite transformations, transformations using coordinates)	6, 8, 15, 37	2 + 2 + 2 + 4 = 10		
14. Locus & Constructions (simple and compound locus, locus using coordinates, compass constructions)	32, 36	2 + 4 = 6		

Topic	Question Numbers	Number of Points	Your Points	Your Percentage
15. Midpoint and Concurrency Theorems (includes joining midpoints of two sides of a triangle, three sides of a triangle, the sides of a quadrilateral, median of a trapezoid; centroid, orthocenter, incenter, and circumcenter)	20, 25	2 + 2 = 4		
16. Circles and Angle Measurement (includes ≅ tangents from same exterior point, radius ⊥ tangent, tangent circles, common tangents, arcs and chords, diameter ⊥ chord, arc length, area of a sector, center-radius equation, applying transformations)	4, 11, 21, 28	2 + 2 + 2 + 2 = 8		
17. Circles and Similar Triangles (includes proofs, segments of intersecting chords, tangent and secant segments)	23	2		
18. Area of Plane Figures (includes area of a regular polygon)	—	—		
19. Measurement of Solids (volume and lateral area; surface area of a sphere; great circle; comparing similar solids)	26, 30	2 + 2 = 4		

Map to Core Curriculum

Content Band	Item Numbers
Geometric Relationships	14, 26, 27, 30
Constructions	2, 32
Locus	25, 36
Informal and Formal Proofs	1, 3, 4, 5, 7, 13, 16, 18, 20, 22, 23, 24, 28, 29, 33, 34, 38
Transformational Geometry	6, 8, 15, 37
Coordinate Geometry	9, 10, 11, 12, 17, 19, 21, 31, 35

How to Convert Your Raw Score to Your Geometry Regents Examination Score

The conversion chart on the following page must be used to determine your final score on the August 2009 Regents Examination in Geometry. To find your final exam score, locate in the column labeled "Raw Score" the total number of points you scored out of a possible 86 points. Since partial credit is allowed in Parts II, III, and IV of the test, you may need to approximate the credit you would receive for a solution that is not completely correct. Then locate in the adjacent column to the right the scale score that corresponds to your raw score. The scale score is your final Geometry Regents Examination score.

Regents Examination in Geometry—August 2009

Chart for Converting Total Test Raw Scores to Final Examination Scores (Scale Scores)

Raw Score	Scale Score	Raw Score	Scale Score	Raw Score	Scale Score	Raw Score	Scale Score
86	100	64	83	42	65	20	35
85	99	63	82	41	63	19	34
84	97	62	81	40	62	18	32
83	96	61	81	39	61	17	30
82	95	60	80	38	60	16	28
81	94	59	79	37	59	15	27
80	93	58	79	36	58	14	25
79	92	57	78	35	57	13	23
78	91	56	77	34	55	12	21
77	91	55	77	33	54	11	19
76	90	54	76	32	53	10	17
75	89	53	75	31	52	9	15
74	88	52	74	30	50	8	14
73	88	51	73	29	49	7	12
72	87	50	72	28	48	6	10
71	87	49	71	27	46	5	8
70	86	48	71	26	45	4	7
69	86	47	70	25	43	3	5
68	85	46	69	24	42	2	3
67	84	45	68	23	40	1	1
66	84	44	67	22	39	0	0
65	83	43	66	21	37		

Examination
June 2010

Geometry

GEOMETRY REFERENCE SHEET

Volume	Cylinder	$V = Bh$, where B is the area of the base
	Pyramid	$V = \frac{1}{3}Bh$, where B is the area of the base
	Right circular cone	$V = \frac{1}{3}Bh$, where B is the area of the base
	Sphere	$V = \frac{4}{3}\pi r^3$

Lateral area (L)	Right circular cylinder	$L = 2\pi rh$
	Right circular cone	$L = \pi r \ell$, where ℓ is the slant height

Surface area	Sphere	$SA = 4\pi r^2$

PART I

Answer all 28 questions in this part. Each correct answer will receive 2 credits. No partial credit will be allowed. For each question, write in the space provided the number preceding the word or expression that best completes the statement or answers the question. [56 credits]

1 In the diagram below of circle O, chord $\overline{AB} \parallel$ chord $\overline{CD}$, and chord $\overline{CD} \parallel$ chord $\overline{EF}$.

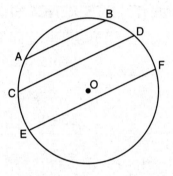

Which statement must be true?

(1) $\overarc{CE} \cong \overarc{DF}$ (3) $\overarc{AC} \cong \overarc{CE}$

(2) $\overarc{AC} \cong \overarc{DF}$ (4) $\overarc{EF} \cong \overarc{CD}$

1 ____1____

2 What is the negation of the statement "I am not going to eat ice cream"?

(1) I like ice cream.
(2) I am going to eat ice cream.
(3) If I eat ice cream, then I like ice cream.
(4) If I don't like ice cream, then I don't eat ice cream.

2 _2_

3 The diagram below shows a right pentagonal prism.

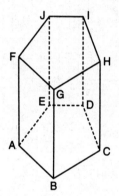

Which statement is always true?

(1) $\overline{BC} \parallel \overline{ED}$ (3) $\overline{FJ} \parallel \overline{IH}$
(2) $\overline{FG} \parallel \overline{CD}$ (4) $\overline{GB} \parallel \overline{HC}$

3 _4_

4 In isosceles triangle ABC, $AB = BC$. Which statement will always be true?

(1) $m\angle B = m\angle A$ (3) $m\angle A = m\angle C$
(2) $m\angle A > m\angle B$ (4) $m\angle C < m\angle B$

4 _3_

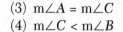

5 The rectangle *ABCD* shown in the diagram below will be reflected across the *x*-axis.

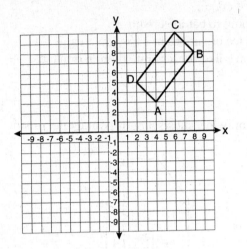

What will *not* be preserved?

(1) slope of $\overline{AB}$
(2) parallelism of $\overline{AB}$ and $\overline{CD}$
(3) length of $\overline{AB}$
(4) measure of $\angle A$

5 ____

6 A right circular cylinder has an altitude of 11 feet and a radius of 5 feet. What is the lateral area, in square feet, of the cylinder, to the *nearest tenth*?

(1) 172.7   (3) 345.4
(2) 172.8   (4) 345.6

6 __4__

7 A transversal intersects two lines. Which condition would always make the two lines parallel?

  (1) Vertical angles are congruent.
  (2) Alternate interior angles are congruent.
  (3) Corresponding angles are supplementary.
  (4) Same-side interior angles are complementary.   7 __2__

8 If the diagonals of a quadrilateral do *not* bisect each other, then the quadrilateral could be a

  (1) rectangle   (3) square
  (2) rhombus   (4) trapezoid   8 __4__

9 What is the converse of the statement "If Bob does his homework, then George gets candy"?

  (1) If George gets candy, then Bob does his homework.
  (2) Bob does his homework if and only if George gets candy.
  (3) If George does not get candy, then Bob does not do his homework.
  (4) If Bob does not do his homework, then George does not get candy.   9 __1__

10 If $\triangle PQR$, $PQ = 8$, $QR = 12$, and $RP = 13$. Which statement about the angles of $\triangle PQR$ must be true?

  (1) $m\angle Q > m\angle P > m\angle R$
  (2) $m\angle Q > m\angle R > m\angle P$
  (3) $m\angle R > m\angle P > m\angle Q$
  (4) $m\angle P > m\angle R > m\angle Q$ 10 __1__

11 Given:

$$y = \frac{1}{4}x - 3$$
$$y = x^2 + 8x + 12$$

In which quadrant will the graphs of the given equations intersect?

(1) I   (3) III
(2) II   (4) IV   11 __3__

12 Which diagram shows the construction of an equilateral triangle?

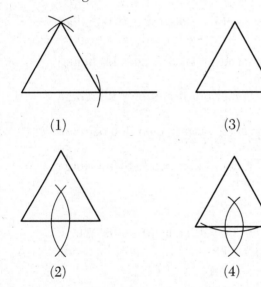

(1)   (3)

(2)   (4)   12 __1__

13 Line segment AB is tangent to circle O at A. Which type of triangle is always formed when points A, B, and O are connected?

(1) right   (3) scalene
(2) obtuse   (4) isosceles

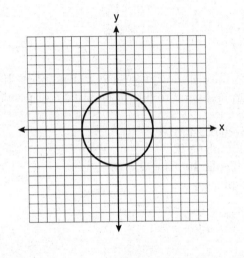

× 13 _4_

14 What is an equation for the circle shown in the graph below?

$= 16$

(1) $x^2 + y^2 = 2$ (3) $x^2 + y^2 = 8$
(2) $x^2 + y^2 = 4$ (4) $x^2 + y^2 = 16$

14 _4_

15 Which transformation can map the letter **S** onto itself?

(1) glide reflection   (3) line reflection
(2) translation   (4) rotation   15 __1__

16 In isosceles trapezoid $ABCD$, $\overline{AB} \cong \overline{CD}$. If $BC = 20$, $AD = 36$, and $AB = 17$, what is the length of the altitude of the trapezoid?

(1) 10   (3) 15
(2) 12   (4) 16   16 __3__

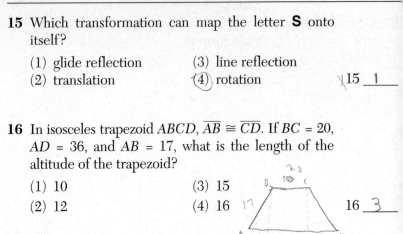

17 In plane $\mathcal{P}$, lines m and n intersect at point A. If line k is perpendicular to line m and line n at point A, then line k is
(1) contained in plane $\mathcal{P}$
(2) parallel to plane $\mathcal{P}$
(3) perpendicular to plane $\mathcal{P}$
(4) skew to plane $\mathcal{P}$ 17 __3__

18 The diagram below shows $\overline{AB}$ and $\overline{DE}$.

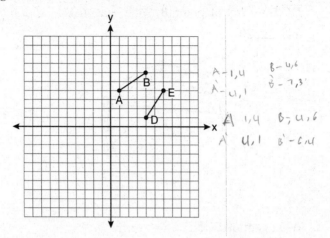

$A - 1, 4 \qquad B - 4, 6$
$A - 4, 1 \qquad B - 7, 3$

$A \ 1, 4 \qquad B, 4, 6$
$A' \ 4, 1 \qquad B' - 6, 4$

Which transformation will move $\overline{AB}$ onto $\overline{DE}$ such that point D is the image of point A and point E is the image of point B?

(1) $T_{3,-3}$ (3) $R_{90°}$

(2) $D_{\frac{1}{2}}$ (4) $r_{y\,=\,x}$ 18 __4__

19 In the diagram below of circle O, chords $\overline{AE}$ and $\overline{DC}$ intersect at point B, such that m$\widehat{AC}$ = 36 and m$\widehat{DE}$ = 20.

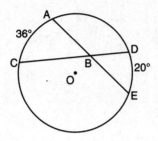

What is m$\angle ABC$?

(1) 56 (3) 28

(2) 36 (4) 8

19 _3_

20 The diagram below shows the construction of a line through point *P* perpendicular to line *m*.

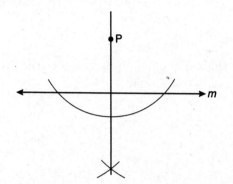

Which statement is demonstrated by this construction?

(1) If a line is parallel to a line that is perpendicular to a third line, then the line is also perpendicular to the third line.

(2) The set of points equidistant from the endpoints of a line segment is the perpendicular bisector of the segment.

(3) Two lines are perpendicular if they are equidistant from a given point.

(4) Two lines are perpendicular if they intersect to form a vertical line.   20 __2__

21 What is the length, to the *nearest tenth*, of the line segment joining the points (−4,2) and (146,52)?

(1) 141.4   (3) 151.9
(2) 150.5   (4) 158.1   21 __4__

$$\sqrt{(52-2)^2 + (146+4)^2}$$
$$50^2 + 150^2$$

22 What is the slope of a line perpendicular to the line whose equation is $y = 3x + 4$?

$y = -\frac{1}{3}x + c$

(1) $\frac{1}{3}$ (3) 3

(2) $-\frac{1}{3}$ (4) −3   22 __2__

23 In the diagram below of circle O, secant $\overline{AB}$ intersects circle O at D, secant $\overline{AOC}$ intersects circle O at E, $AE = 4$, $AB = 12$, and $DB = 6$.

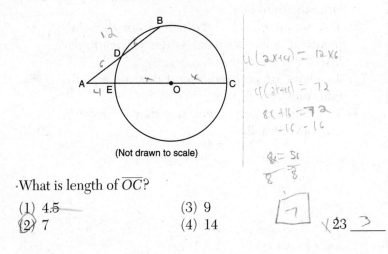

(Not drawn to scale)

$4(2x+4) = 12 \times 6$

$c(2x+4) = 72$

$8x + 16 = 72$
$ -16 \quad -16$

$\dfrac{8x}{8} = \dfrac{56}{8}$

$\boxed{7}$

·What is length of $\overline{OC}$?

(1) 4.5   (3) 9
(2) 7   (4) 14   ✓23 __3__

24 The diagram below shows a pennant in the shape of an isosceles triangle. The equal sides each measure 13, the altitude is $x + 7$, and the base is $2x$.

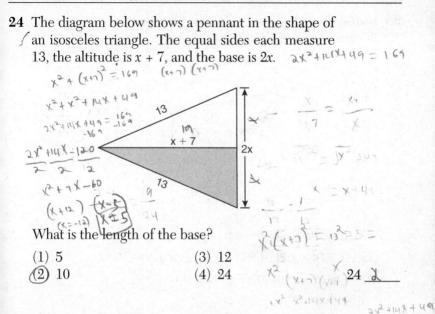

What is the length of the base?

(1) 5   (3) 12

(2) 10   (4) 24

25 In the diagram below of $\triangle ABC$, $\overline{CD}$ is the bisector of $\angle BCA$, $\overline{AE}$ is the bisector of $\angle CAB$, and $\overline{BG}$ is drawn.

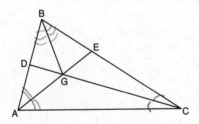

Which statement must be true?

(1) $DG = EG$ (3) $\angle AEB \cong \angle AEC$

(2) $AG = BG$ (4) $\angle DBG \cong \angle EBG$ 25 ___

26 In the diagram below of circle O, chords $\overline{AD}$ and $\overline{BC}$ intersect at E.

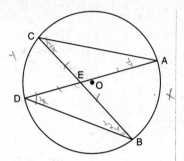

Which relationship must be true?

(1) $\triangle CAE \cong \triangle DBE$ (3) $\angle ACB \cong \angle CBD$

(2) $\triangle AEC \sim \triangle BED$ (4) $\overset{\frown}{CA} \cong \overset{\frown}{DB}$ 26 __2__

27 Two lines are represented by the equations $-\frac{1}{2}y = 6x + 10$ and $y = mx$. For which value of m will the lines be parallel?

(1) -12 (3) 3

(2) -3 (4) 12 27 __1__

$$-\frac{1}{2}y = 6x + 10$$
$$\frac{-\frac{1}{2}}{} \quad \frac{-\frac{1}{2}}{}$$
$$y = -12x + 10$$

28 The coordinates of the vertices of parallelogram $ABCD$ are $A(-3,2)$, $B(-2,-1)$, $C(4,1)$, and $D(3,4)$. The slopes of which line segments could be calculated to show that $ABCD$ is a rectangle?

(1) $\overline{AB}$ and $\overline{DC}$ (3) $\overline{AD}$ and $\overline{BC}$

(2) $\overline{AB}$ and $\overline{BC}$ (4) $\overline{AC}$ and $\overline{BD}$ 28 __2__

PART II

Answer all 6 questions in this part. Each correct answer will receive 2 credits. Clearly indicate the necessary steps, including appropriate formula substitutions, diagrams, graphs, charts, etc. For all questions in this part, a correct numerical answer with no work shown will receive only 1 credit.   [12 credits]

29 Tim is going to paint a wooden sphere that has a diameter of 12 inches. Find the surface area of the sphere, to the *nearest square inch*.

$4\pi r^2$

$4\pi 6^2$

452.

30 In the diagram below of $\triangle ABC$, $\overline{DE}$ is a midsegment of $\triangle ABC$, $DE = 7$, $AB = 10$, and $BC = 13$. Find the perimeter of $\triangle ABC$.

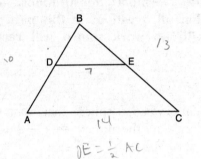

$DE = \frac{1}{2} AC$

31 In right $\triangle DEF$, $m\angle D = 90$ and $m\angle F$ is 12 degrees less than twice $m\angle E$. Find $m\angle E$.

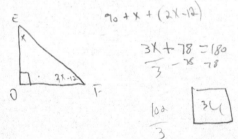

$90 + x + (2x - 12)$

$3x + 78 = 180$
$\frac{3}{3} - 78 \quad 78$

$\frac{102}{3}$ $\boxed{34}$

32 Triangle *XYZ*, shown in the diagram below, is reflected over the line $x = 2$. State the coordinates of $\triangle X'Y'Z'$, the image of $\triangle XYZ$.

X - -1, 1   X' - 5, 1
Y - 0, 4   Y' - 4, 4
Z - 3, 4   Z' - -7, 4

2(2) - X , Y

4+1 , Y

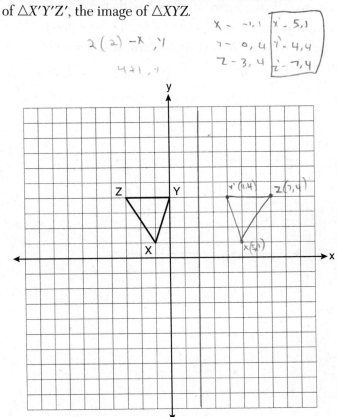

33 Two lines, $\overleftrightarrow{AB}$ and $\overleftrightarrow{CRD}$, are parallel and 10 inches apart. Sketch the locus of all points that are equidistant from $\overleftrightarrow{AB}$ and $\overleftrightarrow{CRD}$ and 7 inches from point R. Label with an **X** each point that satisfies both conditions.

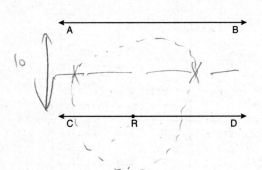

34 The base of a pyramid is a rectangle with a width of 6 cm and a length of 8 cm. Find, in centimeters, the height of the pyramid if the volume is 288 cm^3.

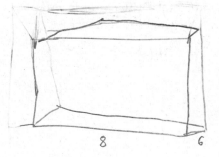

8 6 $\frac{1}{3}$

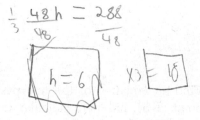

$\frac{1}{3}$ $\dfrac{48h}{48} = \dfrac{288}{48}$

$h = 6$ $\times 3 = 18$

PART III

Answer all 3 questions in this part. Each correct answer will receive 4 credits. Clearly indicate the necessary steps, including appropriate formula substitutions, diagrams, graphs, charts, etc. For all questions in this part, a correct numerical answer with no work shown will receive only 1 credit. [12 credits]

35 Given: Quadrilateral $ABCD$ with $\overline{AB} \cong \overline{CD}$, $\overline{AD} \cong \overline{BC}$, and diagonal $\overline{BD}$ is drawn

Prove: $\angle BDC \cong \angle ABD$

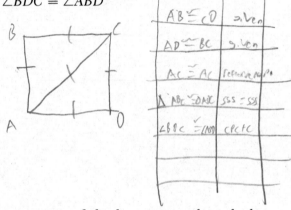

$AB \cong CD$	given
$AD \cong BC$	given
$AC \cong AC$	reflexive prop.
$\triangle ABC \cong \triangle DAC$	SSS ≅ SSS
$\angle BDC \cong \angle ABD$	CPCTC

36 Find an equation of the line passing through the point (6,5) and perpendicular to the line whose equation is $2y + 3x = 6$.

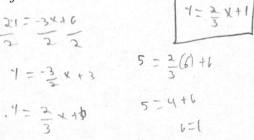

$$\frac{2y}{2} = \frac{-3x}{2} + \frac{6}{2}$$

$$y = \frac{-3}{2}x + 3$$

$$y = \frac{2}{3}x + b$$

$$y = \frac{2}{3}x + 1$$

$$5 = \frac{2}{3}(6) + b$$

$$5 = 4 + b$$

$$b = 1$$

37 Write an equation of the circle whose diameter $\overline{AB}$ has endpoints $A(-4,2)$ and $B(4,-4)$. [The use of the grid below is optional.]

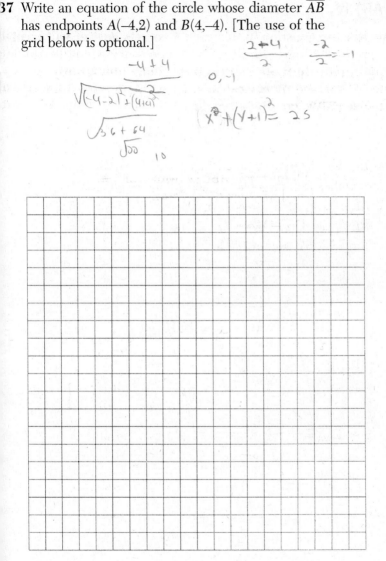

$$\frac{2+4}{2} \quad \frac{-2}{2} = -1$$

$$\frac{-4+4}{2}$$

$$\sqrt{(-4-2)^2+(4+4)^2}$$

$$0, -1$$

$$\sqrt{36+64}$$

$$\sqrt{100} \quad 10$$

$$(x^2+(y+1)^2 = 25$$

PART IV

Answer the question in this part. A correct answer will receive 6 credits. Clearly indicate the necessary steps, including appropriate formula substitutions, diagrams, graphs, charts, etc. A correct numerical answer with no work shown will receive only 1 credit.   [6 credits]

38 In the diagram below, quadrilateral *STAR* is a rhombus with diagonals $\overline{SA}$ and $\overline{TR}$ intersecting at *E*.
$ST = 3x + 30$, $SR = 8x - 5$, $SE = 3z$, $TE = 5z + 5$,
$AE = 4z - 8$, m$\angle RTA = 5y - 2$, and m$\angle TAS = 9y + 8$.
Find *SR*, *RT*, and m$\angle TAS$.

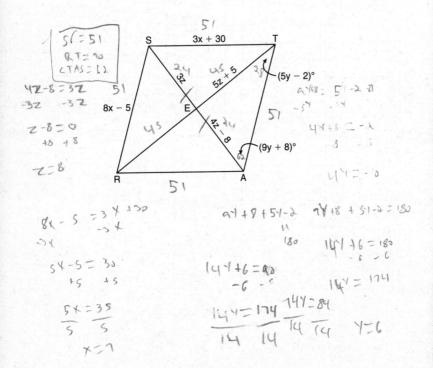

Answers
June 2010
Geometry

Answer Key

PART I

1. (1)	**8.** (4)	**15.** (4)	**22.** (2)
2. (2)	**9.** (1)	**16.** (3)	**23.** (2)
3. (4)	**10.** (1)	**17.** (3)	**24.** (2)
4. (3)	**11.** (3)	**18.** (4)	**25.** (4)
5. (1)	**12.** (1)	**19.** (3)	**26.** (2)
6. (4)	**13.** (1)	**20.** (2)	**27.** (1)
7. (2)	**14.** (4)	**21.** (4)	**28.** (2)

PART II

29. 452
30. 37
31. 34
32. $X'(5,1)$, $Y'(4,4)$, and $Z'(7, 4)$
33. See *Answers Explained*.
34. 18

PART III

35. See *Answers Explained*.

36. $y - 5 = \frac{2}{3}(x - 6)$ or

$y = \frac{2}{3}x + 1$

37. $x^2 + (y + 1)^2 = 25$

PART IV

38. $SR = 51$, $RT = 90$, and m$\angle TAS = 62$

In **PARTS II–IV** you are required to show how you arrived at your answers. For sample methods of solutions, see the *Answers Explained* section.

Answers Explained

PART I

1. In the accompanying diagram of circle O, it is given that chord $\overline{AB}$ ‖ chord $\overline{CD}$ and that chord $\overline{CD}$ ‖ chord $\overline{EF}$.

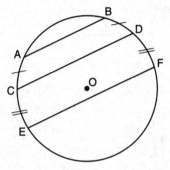

Since parallel chords cut off congruent arcs as indicated in the diagram, it must be true that $\overarc{AC} \cong \overarc{BD}$ and $\overarc{CE} \cong \overarc{DF}$.

The correct choice is **(1)**.

2. A statement and its negation have opposite truth values. The given statement is "I am not going to eat ice cream." The statement with the opposite truth value is "I am going to eat ice cream."

The correct choice is **(2)**.

3. It is given that the accompanying diagram represents a right pentago-nal prism.

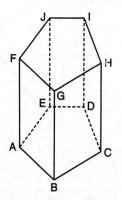

Since the lateral faces of a right prism are rectangles, quadrilateral *GBCH* is a rectangle. Because opposite sides of a rectangle are parallel, $\overline{GB} \parallel \overline{HC}$.

The correct choice is **(4)**.

4. If two sides of a triangle have the same lengths, the angles opposite these sides have the same measures. It is given that in isosceles triangle *ABC*, *AB = BC*. Hence, m∠*A* = m∠*C*.

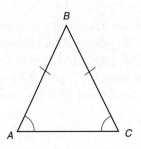

The correct choice is **(3)**.

5. In the accompanying diagram, rectangle $ABCD$ is reflected across the x-axis. You are asked to identify from among the answer choices a property that will not be preserved under a reflection.

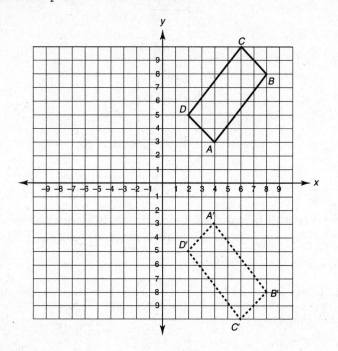

A reflection of a figure across a line produces its mirror image. As a result angle measure and distance are preserved but not orientation. Hence, reflecting $ABCD$ across the x-axis preserves parallelism of $\overline{AB}$ and $\overline{CD}$, the length of $\overline{AB}$, and the measure of $\angle A$. Since the slope of $\overline{AB}$ is positive and the slope of its image, $\overline{A'B'}$, is negative, the slope of $\overline{AB}$ is not preserved.

The correct choice is **(1)**.

6. The lateral area, L, of a right circular cylinder is given by the formula $L = 2\pi rh$, where r is the radius of the circular base and h is the altitude. Since it is given that $r = 5$ feet and $h = 11$ feet:

$$L = 2\pi \times 5 \times 11$$

Use your calculator and the stored value of π: $\approx 345.5751919.$

The lateral area of the cylinder to the *nearest tenth* is 345.6.

The correct choice is **(4)**.

7. If a transversal intersects two lines, as shown in the accompanying diagram, the lines are parallel when alternate interior angles 1 and 2 (or 3 and 4) are congruent.

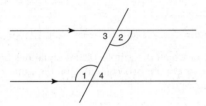

The correct choice is **(2)**.

8. The diagonals of a parallelogram bisect each other. Since a rectangle, rhombus, and square are special parallelograms, the diagonals in each of these figures bisect each other. The diagonals of a trapezoid do not bisect each other.

The correct choice is **(4)**.

9. The converse of a conditional statement of the form "If p, then q" is a statement of the form "If q, then p." To form the converse of a conditional statement, interchange the if and then clauses:

- Given statement: If <u>Bob does his homework</u>, then <u>George gets candy</u>.
- Converse: If <u>George gets candy</u>, then <u>Bob does his homework</u>.

The correct choice is **(1)**.

10. When comparing the measures of the angles of a triangle, the larger angle lies opposite the longer side. It is given that in $\triangle PQR$, $PQ = 8$, $QR = 12$, and $RP = 13$, as shown in the accompanying figure. Since $RP > QR > PQ$, the angles opposite these sides have the same size relationship. Thus $m\angle Q > m\angle P > m\angle R$.

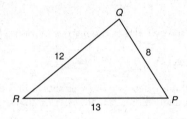

The correct choice is **(1)**.

11. Given that $y = \dfrac{1}{4}x - 3$ and $y = x^2 + 8x + 12$, you are asked to determine the quadrant in which the graphs of the equations intersect. Using your graphing calculator, graph the two equations on the same set of axes as shown in the accompanying figure.

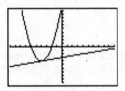

The graphs intersect in quadrant III.

The correct choice is **(3)**.

12. The accompanying diagram shows horizontal segment $\overline{AB}$ marked off using point B as the center.

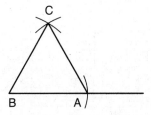

The intersecting arcs at point C are constructed by keeping the same compass setting and drawing arcs using points A and B as centers, thereby constructing an equilateral triangle with $AB = BC = AC$.

The correct choice is (**1**).

13. A radius drawn to a tangent at the point of contact with the circle is perpendicular to the tangent. In the accompanying diagram of circle O, line segment AB is tangent to circle O at A. When points A, B, and O are connected, as illustrated in the accompanying diagram, $\angle OAB$ is a right angle. So the triangle formed is always a right triangle.

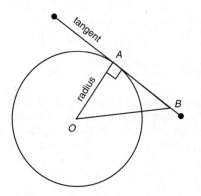

The correct choice is (**1**).

14. The general form of an equation of a circle with center at the origin and radius r is $x^2 + y^2 = r^2$. You are asked for the equation of the circle in the accompanying graph.

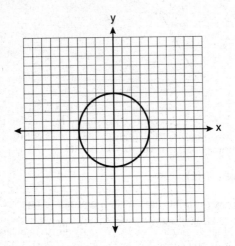

As no scale is indicated, you can assume for this exam that each grid box represents 1 unit. Because the circle is centered at the origin and intersects the x-axis at $(4,0)$, $r = 4$ units. Hence, an equation of the given circle is $x^2 + y^2 = 4^2$ or, equivalently, $x^2 + y^2 = 16$.

The correct choice is (**4**).

15. A transformation maps a figure onto itself if the image exactly coincides with the original figure. If the page on which the letter **S** is written is turned upside down, the letter will look exactly the same. Since turning a page upside down corresponds to a rotation of $180°$, a rotation of $180°$ can map the letter **S** onto itself.

The correct choice is (**4**).

16. You are asked to find the length of the altitude of isosceles trapezoid *ABCD* with $\overline{AB} \cong \overline{CD}$, *BC* = 20, *AD* = 36, and *AB* = 17, as shown in the accompanying diagram.

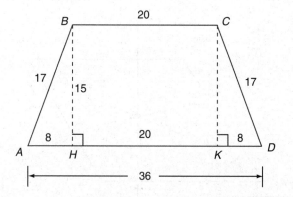

- Dropping altitudes $\overline{BH}$ and $\overline{CK}$ to $\overline{AD}$ forms rectangle *BHKC* with *HK* = *BC* = 20.

- Right △*AHB* is congruent to right △*DKC*, so *AH* = *DK*. Since *AH* + *DK* = 36 − 20 = 16, *AH* = *DK* = $\frac{1}{2} \times 16 = 8$.

- The lengths of the sides of right △*AHB* form an 8-15-17 Pythagorean triple in which *BH* = 15. If you did not recognize this Pythagorean triple, use the Pythagorean theorem to find *BH*:

$$8^2 + (BH)^2 = 17^2$$
$$(BH)^2 = 289 - 64$$
$$\sqrt{(BH)^2} = \sqrt{225}$$
$$BH = 15$$

The correct choice is **(3)**.

17. It is given that in plane $\mathcal{P}$, lines m and n intersect at point A and that line k is perpendicular to line m and line n at point A, as shown in the accompanying diagram.

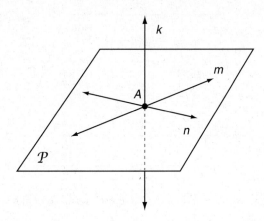

If $k \perp m$ and $k \perp n$ at point A, then $k \perp$ plane $\mathcal{P}$.

If a line is perpendicular to intersecting lines at their point of intersection, the line is perpendicular to the plane determined by them. Hence, line k is perpendicular to plane $\mathcal{P}$.

The correct choice is **(3)**.

18. You are asked to determine the transformation that will map $\overline{AB}$ onto $\overline{DE}$ such that D is the image of point A and E is the image of point B as shown in the accompanying diagram.

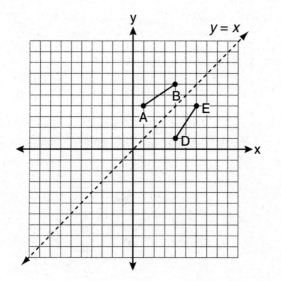

Examine each choice in turn.

- Choice (1): The notation $T_{3,-3}$ represents a translation or shift of $\overline{AB}$ 3 units horizontally to the right and 3 units vertically down. This does not map point B onto point E. ✗

- Choice (2): The notation $D_{\frac{1}{2}}$ represents a dilation or size transformation using a scale factor of $\frac{1}{2}$. Since multiplying the coordinates of the endpoints of $\overline{AB}$ by $\frac{1}{2}$ does not produce the corresponding endpoints of $\overline{DE}$, $D_{\frac{1}{2}}$ will not map $\overline{AB}$ onto $\overline{DE}$. ✗

- Choice (3): The notation $R_{90°}$ represents a rotation of 90°. Rotating $\overline{AB}$ 90° in either the clockwise or counterclockwise directions does not produce $\overline{DE}$. ✗

- Choice (4): The notation $r_{y=x}$ represents a reflection in the line $y = x$. Draw the line $y = x$ as shown in the accompanying diagram. Since D is the image of point A and E is the image of point B under a reflection in the line $y = x$, this is the correct choice. ✔

The correct choice is **(4)**.

19. It is given that in the accompanying diagram of circle O, chords $\overline{AE}$ and $\overline{DC}$ intersect at point B with m $\widehat{AC}$ = 36 and m $\widehat{DE}$ = 20. You are asked to find m$\angle ABC$.

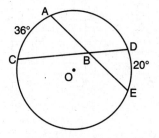

The measure of an angle formed by two chords intersecting in the interior of a circle is one-half the sum of the measures of the intercepted arcs.

$$m\angle ABC = \frac{1}{2}\left(m\widehat{AC} + m\widehat{DE}\right)$$

$$= \frac{1}{2}\left(36 + 20\right)$$

$$= \frac{1}{2}(56)$$

$$= 28$$

The correct choice is **(3)**.

20. It is given that the diagram below shows the construction of a line through point P perpendicular to line m. You are asked to identify from among the answer choices the statement demonstrated by the construction.

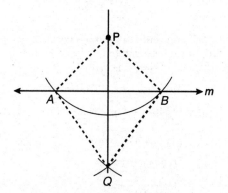

Label the endpoints of the line segment determined by the arc as A and B.

- Since $\overline{PA}$ and $\overline{PB}$ are radii of a circle with center at P, $PA = PB$. Thus, point P is equidistant from points A and B.

- Call the point at which the two arcs intersect below line m point Q. Since the same compass setting was used to draw the two arcs with A and B as centers, $AQ = BQ$ so that point Q is also equidistant from points A and B.

- The constructed perpendicular is the line determined by two points, P and Q, that are each equidistant from the endpoints of $\overline{AB}$. This suggests that the set of all points equidistant from the endpoints of a line segment is the perpendicular bisector of that segment.

The correct choice is **(2)**.

21. To find the length, d, of the line segment joining the given points $(-4,2)$ and $(146,52)$, use the distance formula:

$$d = \sqrt{(\Delta x)^2 + (\Delta y)^2}$$

In this formula, Δx represents the difference in the x-coordinates of the two given points and Δy represents the difference in their y-coordinates, with the subtractions performed in the same order:

$$
\begin{aligned}
d &= \sqrt{\left(146-(-4)\right)^2 + (52-2)^2} \\
&= \sqrt{(146+4)^2 + (50)^2} \\
&= \sqrt{22,500 + 2,500} \\
&= \sqrt{25,000} \\
&\approx 158.113883
\end{aligned}
$$

The length of the line segment to the *nearest tenth* is 158.1.

The correct choice is (**4**).

22. Perpendicular lines have slopes that are negative reciprocals. To find the slope of a line perpendicular to the line $y = 3x + 4$, find the negative reciprocal of the slope of the given line.

In general, a line whose equation is in the slope-intercept form $y = mx + b$ has a slope of m and a y-intercept of b.

For the equation $y = 3x + 4$, $m = 3$ and $b = 4$. Hence, the slope of the given line is 3.

Since the negative reciprocal of 3 is $-\dfrac{1}{3}$, a line perpendicular to the line $y = 3x + 4$ has a slope of $-\dfrac{1}{3}$.

The correct choice is (**2**).

23. In the accompanying diagram of circle O, secant $\overline{AB}$ intersects circle O at D, secant $\overline{AOC}$ intersects circle O at E, $AE = 4$, $AB = 12$, and $DB = 6$. You are asked to find the length of $\overline{OC}$.

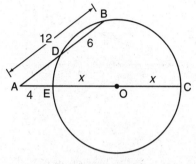

(Not drawn to scale)

$\overline{OC}$ is a radius of circle. If x represents the length of $\overline{OC}$, the length of diameter $\overline{COE}$ can be represented by $2x$.

If two secants are drawn to a circle from the same exterior point, the product of the lengths of the first secant and its external segment is equal to the product of the lengths of the second secant and its external segment:

$$AE \times AC = AD \times AB$$

Since $AD = AB - DB = 12 - 6 = 6$:

$$\underbrace{4}_{AE} \times \underbrace{(2x+4)}_{AC} = \underbrace{6}_{AD} \times \underbrace{12}_{AB}$$

$$8x + 16 = 72$$

$$8x = 72 - 16$$

$$\frac{8x}{8} = \frac{56}{8}$$

$$x = 7$$

The length of $\overline{OC}$ is 7.

The correct choice is **(2)**.

24. It is given that the accompanying diagram shows a pennant in the shape of an isosceles triangle such that the equal sides each measure 13, the altitude is $x + 7$, and the base is $2x$. You are asked to find the length of the base.

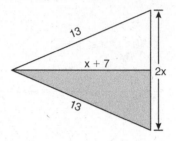

Since the altitude drawn to the base of an isosceles triangle bisects the base, the altitude forms two congruent right triangles with side lengths of x, $x + 7$, and 13.

Use the Pythagorean theorem to find x:

$$x^2 + (x+7)^2 = 13^2$$

$$x^2 + (x^2 + 14x + 49) = 169$$

$$2x^2 + 14x + 49 - 169 = 0$$

$$\frac{2x^2}{2} + \frac{14x}{2} - \frac{120}{2} = 0$$

$$x^2 + 7x - 60 = 0$$

Solve by factoring: $(x - 5)(x + 12) = 0$

Set each factor equal to 0: $x - 5 = 0$ or $x + 12 = 0$

$x = 5$ or $x = -12$

Reject $x = -12$ as a solution since the length of a side of a triangle must be greater than 0.

Since $x = 5$, the length of the base $= 2x = 2(5) = 10$.

The correct choice is **(2)**.

25. It is given that in the accompanying diagram of △*ABC*, $\overline{CD}$ is the bisector of ∠*BCA*, $\overline{AE}$ is the bisector of ∠*CAB*, and $\overline{BG}$ is drawn.

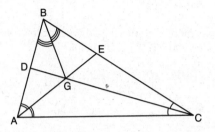

The bisectors of the three angles of any triangle are concurrent at a point. Because $\overline{CD}$, $\overline{AE}$, and $\overline{BG}$ meet at point *G*, $\overline{BG}$ must also be an angle bisector. Since an angle bisector divides an angle into two congruent angles, ∠*DBG* ≅ ∠*EBG*.

The correct choice is (**4**).

26. In the accompanying diagram of circle *O*, chords $\overline{AD}$ and $\overline{BC}$ intersect at *E*.

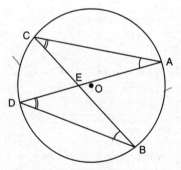

In a circle, inscribed angles that intercept the same arc are congruent. Since angles *A* and *B* each intercept the same arc, $\overset{\frown}{CD}$, ∠*A* ≅ ∠*B*. Because angles *C* and *D* each intercept the same arc, $\overset{\frown}{AB}$, ∠*C* ≅ ∠*D*. By the AA Theorem of Similarity, △*AEC* ~ △*BED*.

The correct choice is (**2**).

27. You are asked to find the value of m for which the lines represented by the equations $-\frac{1}{2}y = 6x + 10$ and $y = mx$ are parallel. Two lines are parallel when they have the same slope.

Rewrite the first equation in the slope-intercept form $y = mx + b$, where m is the slope of the line and b is the y-intercept.

$$-2\left(-\frac{1}{2}y\right) = -2(6x + 10)$$

$$y = -12x - 20$$

The slope of the line $y = -12x - 20$ is -12, the coefficient of x. Since the slope of the line $y = mx$ is m, the given lines will be parallel when $m = -12$.

The correct choice is (**1**).

28. The coordinates of the vertices of parallelogram $ABCD$ are $A(-3,2)$, $B(-2,-1)$, $C(4,1)$, and $D(3,4)$. You need to identify the pair of segments whose slopes could be calculated to show that $ABCD$ is a rectangle. Sketch the rectangle:

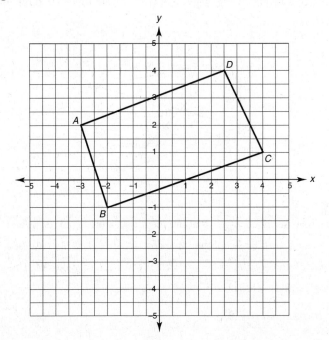

A parallelogram is a rectangle if a pair of adjacent sides have slopes that are negative reciprocals and, as a result, form a right angle. Examine the answer choices until you find a pair of adjacent sides, such as $\overline{AB}$ and $\overline{BC}$.

The correct choice is **(2)**.

PART II

29. To find the surface area, SA, of a sphere that has a diameter of 12 inches, use the formula $SA = 4\pi r^2$, where r is the radius of the sphere. Since the diameter of the sphere is 12 inches, the radius is 6 inches. Use a calculator and the stored value of π:

$$SA = 4\pi \cdot 6^2$$

$$= 144\pi$$

$$\approx 452.3893421$$

The surface area of the sphere to the *nearest square inch* is **452 in²**.

30. It is given that in the accompanying diagram of $\triangle ABC$, $\overline{DE}$ is a mid-segment of $\triangle ABC$, $DE = 7$, $AB = 10$, and $BC = 13$.

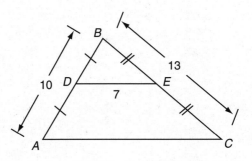

Since a midsegment of a triangle joins the midpoints of two sides of a triangle, points D and E are the midpoints of $\overline{AB}$ and $\overline{BC}$, respectively. A line segment that joins the midpoints of two sides of a triangle is parallel to the third side and is one-half of its length.

$$DE = \frac{1}{2}AC$$

$$7 = \frac{1}{2}AC$$

$$AC = 2 \times 7$$

$$= 14$$

The perimeter of $\triangle ABC$ is the sum of the lengths of its sides:

$$\text{Perimeter} = 10 + 13 + 14$$
$$= 37$$

The perimeter of $\triangle ABC$ is **37**.

31. It is given that in right $\triangle DEF$, $m\angle D = 90$ and $m\angle F$ is 12 less than twice $m\angle E$. If x represents $m\angle E$, then $2x - 12$ represents $m\angle F$. Since the sum of the measures of three angles of a triangle is 180:

$$(2x - 12) + x + 90 = 180$$
$$3x + 78 = 180$$
$$3x = 180 - 78$$
$$\frac{3x}{3} = \frac{102}{3}$$
$$x = 34$$

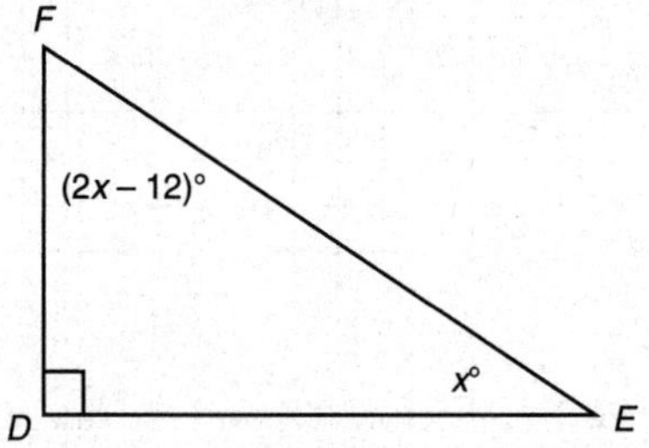

Thus $m\angle E = $ **34**.

32. The accompanying diagram shows $\triangle XYZ$. You are asked to determine the coordinates of $\triangle X'Y'Z'$, the image of $\triangle XYZ$ after a reflection over the line $x = 2$.

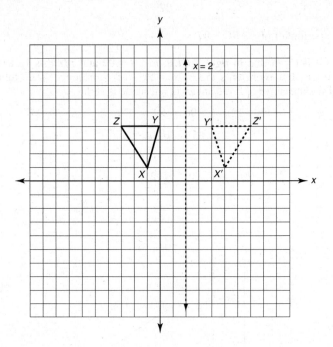

Draw the line $x = 2$. To reflect $\triangle XYZ$ over the line $x = 2$, reflect points X Y, and Z across the line of reflection. First count the number of unit boxes point X is located horizontally from the line $x = 2$. Then locate point X', the image of point X, by finding the point on the opposite side of the line $x = 2$ and the same number of units from it. Do the same for points Y and Z.

Since point X is 3 units to the left of the line of reflection, its image will be 3 units to the right of the line $x = 2$:

$$X(-1,1) \rightarrow X'(2+3,1)$$

Since point Y is 2 units to the left of the line of reflection, its image will be 2 units to the right of the line $x = 2$:

$$Y(0,4) \rightarrow Y'(2+2,4)$$

Because point Z is 5 units to the left of the line of reflection, its image will be 5 units to the right of the line $x = 2$:

$$Z(-3,4) \rightarrow Z'(2+5,4)$$

Thus, the coordinates of the vertices of $\triangle X'Y'Z'$ are **X'(5,1)**, **Y'(4,4)**, and **Z'(7,4)**.

33. It is given that in the accompanying diagram $\overleftrightarrow{AB}$ and $\overleftrightarrow{CRD}$ are parallel and 10 inches apart. You are asked to sketch the locus of all points that are equidistant from $\overleftrightarrow{AB}$ and $\overleftrightarrow{CRD}$ and 7 inches from point R. You must label with an **X** each point that satisfies both conditions.

Sketch the locus of all points that are equidistant from $\overleftrightarrow{AB}$ and $\overleftrightarrow{CRD}$ by drawing a line parallel to the given lines and midway between them. This new line will be 5 inches from each line.

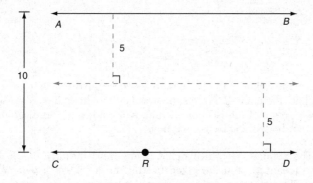

Sketch the set of all points that are 7 units from point R by drawing a circle with its center at R and a radius of 7 units.

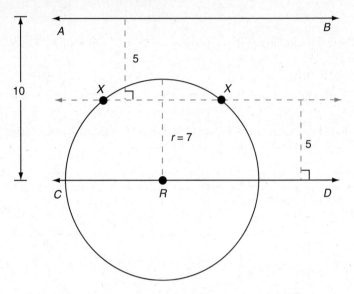

Because the radius of the circle is greater than the distance from point R to the new line, the circle intersects the new line in exactly 2 points. Mark each of the two points of intersection with an **X** as shown in the diagram. These points satisfy both of the locus conditions.

34. The volume, V, of a pyramid is one-third the area of the base, B, times the height, h. You are asked to find the height of a pyramid whose volume is 288 cm^3 given that its rectangular base has a width of 6 cm and a length of 8 cm.

$$V = \frac{1}{3}Bh$$

$$288 \text{ cm}^3 = \frac{1}{3} \times \overbrace{(8 \text{ cm} \times 6 \text{ cm})}^{\text{Area of rectangular base}} \times h$$

$$288 = 16h$$

$$\frac{16h}{16} = \frac{288}{16}$$

$$h = 18 \text{ cm}$$

The height of the pyramid is **18 cm**.

PART III

35. In the accompanying diagram, you are given quadrilateral *ABCD* with $\overline{AB} \cong \overline{CD}$, $\overline{AD} \cong \overline{BC}$, and diagonal $\overline{BD}$ is drawn. You have to prove that $\angle BDC \cong \angle ABD$.

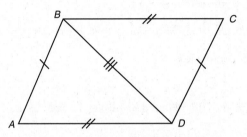

Plan: Prove $\triangle ABD \cong \triangle CDB$ by SSS $\cong$ SSS.

Statement	Reason
1. Quadrilateral *ABCD* with $\overline{AB} \cong \overline{CD}$, $\overline{AD} \cong \overline{BC}$, and diagonal $\overline{BD}$.	1. Given.
2. $\overline{BD} \cong \overline{BD}$.	2. Reflexive property of congruence.
3. $\triangle ABD \cong \triangle CDB$.	3. SSS $\cong$ SSS
4. $\angle BDC \cong \angle ABD$.	4. Corresponding parts of congruent triangles are congruent (CPCTC).

36. You are asked to find an equation of the line passing through the point (6,5) and perpendicular to the line whose equation is $2y + 3x = 6$.

Find the slope of the given line by writing its equation in slope-intercept form, $y = mx + b$. If $2y + 3x = 6$, then $2y = -3x + 6$ so $y = -\frac{3}{2}x + 3$. The slope of the given line is $-\frac{3}{2}$, the coefficient of x.

Since perpendicular lines have slopes that are negative reciprocals, the slope, m, of the required line is $\frac{3}{2}$.

Method 1:

Write an equation of the required line using the point-slope form $y - y_1 = m(x - x_1)$, where $(x_1, y_1) = (6,5)$ and $m = \frac{2}{3}$:

$$y - 5 = \frac{2}{3}(x - 6)$$

Method 2:

Write an equation of the required line using the slope-intercept form $y = mx + b$, where $m = \frac{2}{3}$ and b is determined by substituting the coordinates (6,5) into the equation $y = mx + b$:

$$5 = \frac{2}{\cancel{3}}(\cancel{6}) + b$$

$$5 = 4 + b$$

$$1 = b$$

Because $m = \frac{2}{3}$ and $b = 1$, an equation of the required line is $y = \frac{2}{3}x + 1$.

The equation of the line can be written as $\boldsymbol{y - 5 = \frac{2}{3}(x - 6)}$ or $\boldsymbol{y = \frac{2}{3}x + 1}$.

37. You are required to write an equation of the circle that has diameter $\overline{AB}$ with endpoints $A(-4,2)$ and $B(4,-4)$. To write an equation of the circle, you need to know its center and radius. The center of the circle is the midpoint of a diameter, and the radius is the distance from the center of the circle to a point on the circle.

To find the coordinates $O(x,y)$ of the center of the circle, use the midpoint formula by taking the averages of the corresponding coordinates of the endpoints of diameter $\overline{AB}$:

$$O(x,y) = \left(\frac{-4+4}{2}, \frac{2+(-4)}{2} \right)$$
$$= (0,-1)$$

To find the radius, use the distance formula, $d = \sqrt{(\Delta x)^2 + (\Delta y)^2}$, where Δx represents the difference in the x-coordinates of the center $(0,-1)$ and any point on the circle, such as $A(-4,2)$, and Δy represents the difference in the y-coordinates of the same two points taken in the same order:

$$OA = \sqrt{(0-(-4))^2 + (-1-2)^2}$$
$$= \sqrt{4^2 + (-3)^2}$$
$$= \sqrt{16+9}$$
$$= \sqrt{25}$$
$$= 5$$

The general equation of a circle has the form $(x-h)^2 + (y-k)^2 = r^2$, where (h,k) is the center of the circle and r is the radius. Write an equation of the circle by substituting $(0,-1)$ for (h,k) and 5 for r:

$$(x-0)^2 + (y-(-1))^2 = 5^2$$

An equation of the circle is $x^2 + (y+1)^2 = 25$.

PART IV

38. In the accompanying diagram of rhombus *STAR*, diagonals $\overline{SA}$ and $\overline{TR}$ intersect at *E*. It is also given that $ST = 3x + 30$, $SR = 8x - 5$, $SE = 3z$, $TE = 5z + 5$, $AE = 4z - 8$, m$\angle RTA = 5y - 2$, and m$\angle TAS = 9y + 8$. You are asked to find SR, RT, and m$\angle TAS$.

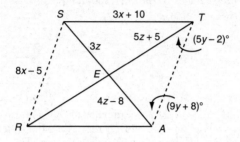

Find *SR*. Since adjacent sides of a rhombus are equal in length:

$$SR = ST$$
$$8x - 5 = 3x + 30$$
$$8x - 3x = 30 + 5$$
$$5x = 35$$
$$\frac{5x}{5} = \frac{35}{5}$$
$$x = 7$$

Substitute 7 for *x* to find *SR*:

$$SR = 8x - 5$$
$$= 8(7) - 5$$
$$= 56 - 5$$
$$= 51$$

Find RT. Since the diagonals of a rhombus bisect each other:

$$AE = SE$$
$$4z - 8 = 3z$$
$$4z - 3z = 8$$
$$z = 8$$

Substitute 8 for z to find TE:

$$TE = 5z + 5$$
$$= 5 \cdot 8 + 5$$
$$= 45$$

Because $TE = \frac{1}{2}RT$, $RT = 2 \times 45 = 90$.

Find m$\angle TAS$. The diagonals of a rhombus intersect at right angles, so m$\angle TEA = 90$. Since the measures of the acute angles of right $\triangle TEA$ add up to 90:

$$\text{m}\angle RTA + \text{m}\angle TAS = 90$$
$$(5y - 2) + (9y + 8) = 90$$
$$14y + 6 = 90$$
$$14y = 90 - 6$$
$$\frac{14y}{14} = \frac{84}{14}$$
$$y = 6$$

Substitute 6 for y to find m$\angle TAS$:

$$\text{m}\angle TAS = 9y + 8$$
$$= 9 \cdot 6 + 8$$
$$= 54 + 8$$
$$= 62$$

$SR = 51$, $RT = 90$, and m$\angle TAS = 62$.

Topic	Question Numbers	Number of Points	Your Points	Your Percentage
1. Logic (negation, conjunction, disjunction, related conditionals, biconditional, logically equivalent statements, logical inference)	2, 9	2 + 2 = 4		
2. Angle & Line Relationships (vertical angles, midpoint, altitude, median, bisector, supplementary angles, complementary angles, parallel and perpendicular lines)	7	2		
3. Parallel & Perpendicular Planes	17	2		
4. Angles of a Triangle & Polygon (sum of angles, exterior angle, base angles theorem, angles of a regular polygon)	4, 31	2 + 2 = 4		
5. Triangle Inequalities (side length restrictions, unequal angles opposite unequal sides, exterior angle)	10	2		
6. Trapezoids & Parallelograms (includes properties of rectangle, rhombus, square, and isosceles trapezoid)	8, 38	2 + 6 = 8		
7. Proofs Involving Congruent Triangles	35	4		
8. Indirect Proof & Mathematical Reasoning	—	—		
9. Ratio & Proportion (includes similar polygons, proportions, proofs involving similar triangles, comparing linear dimensions, and areas of similar triangles)	—	—		
10. Proportions formed by altitude to hypotenuse of right triangle; Pythagorean theorem	24	2		
11. Coordinate Geometry (slopes of parallel and perpendicular lines; slope-intercept and point-slope equations of a line; applications requiring midpoint, distance, or slope formulas; coordinate proofs)	21, 22, 27, 28, 36, 37	2 + 2 + 2 + 2 + 4 + 4 = 16		
12. Solving a Linear-Quadratic System of Equations Graphically	11	2		
13. Transformation Geometry (includes isometries, dilation, symmetry, composite transformations, transformations using coordinates)	5, 15, 18, 32	2 + 2 + 2 + 2 = 8		
14. Locus & Constructions (simple and compound locus, locus using coordinates, compass constructions)	12, 20, 33	2 + 2 + 2 = 6		

Topic	Question Numbers	Number of Points	Your Points	Your Percentage
15. Midpoint and Concurrency Theorems (includes joining midpoints of two sides of a triangle, three sides of a triangle, the sides of a quadrilateral, median of a trapezoid; centroid, orthocenter, incenter, and circumcenter)	25, 30	2 + 2 = 4		
16. Circles and Angle Measurement (includes ≅ tangents from same exterior point, radius ⊥ tangent, tangent circles, common tangents, arcs and chords, diameter ⊥ chord, arc length, area of a sector, center-radius equation, applying transformations)	1, 13, 14, 19	2 + 2 + 2 + 2 = 8		
17. Circles and Similar Triangles (includes proofs, segments of intersecting chords, tangent and secant segments)	23, 26	2 + 2 = 4		
18. Area of Plane Figures (includes area of a regular polygon)	16	2		
19. Measurement of Solids (volume and lateral area; surface area of a sphere; great circle; comparing similar solids)	3, 6, 29, 34	2 + 2 + 2 + 2 = 8		

Map to Core Curriculum

Content Band	Item Numbers
Geometric Relationships	3, 6, 17, 29, 34
Constructions	12, 20
Locus	25, 33
Informal and Formal Proofs	1, 2, 4, 7, 8, 9, 10, 13, 16, 19, 23, 24, 26, 30, 31, 35, 38
Transformational Geometry	5, 15, 18, 32
Coordinate Geometry	11, 14, 21, 22, 27, 28, 36, 37

How to Convert Your Raw Score to Your Geometry Regents Examination Score

The conversion chart on the following page must be used to determine your final score on the June 2010 Regents Examination in Geometry. To find your final exam score, locate in the column labeled "Raw Score" the total number of points you scored out of a possible 86 points. Since partial credit is allowed in Parts II, III, and IV of the test, you may need to approximate the credit you would receive for a solution that is not completely correct. Then locate in the adjacent column to the right the scale score that corresponds to your raw score. The scale score is your final Geometry Regents Examination score.

Regents Examination in Geometry—June 2010

Chart for Converting Total Test Raw Scores to Final Examination Scores (Scale Scores)

Raw Score	Scale Score	Raw Score	Scale Score	Raw Score	Scale Score	Raw Score	Scale Score
86	100	64	80	42	66	20	41
85	98	63	80	41	65	19	40
84	97	62	79	40	64	18	38
83	96	61	79	39	64	17	36
82	95	60	78	38	63	16	34
81	93	59	77	37	62	15	33
80	92	58	77	36	61	14	31
79	91	57	76	35	60	13	29
78	90	56	76	34	59	12	27
77	89	55	75	33	58	11	25
76	88	54	75	32	57	10	22
75	88	53	74	31	56	9	20
74	87	52	73	30	55	8	18
73	86	51	73	29	54	7	16
72	86	50	72	28	52	6	14
71	85	49	71	27	51	5	11
70	84	48	71	26	50	4	9
69	83	47	70	25	49	3	7
68	83	46	69	24	47	2	5
67	82	45	68	23	46	1	2
66	82	44	68	22	44	0	0
65	81	43	67	21	43		

Examination
August 2010
Geometry

GEOMETRY REFERENCE SHEET

Volume	Cylinder	$V = Bh,$ where B is the area of the base
	Pyramid	$V = \frac{1}{3}Bh,$ where B is the area of the base
	Right circular cone	$V = \frac{1}{3}Bh,$ where B is the area of the base
	Sphere	$V = \frac{4}{3}\pi r^3$

Lateral area (L)	Right circular cylinder	$L = 2\pi rh$
	Right circular cone	$L = \pi r\ell,$ where ℓ is the slant height

Surface area	Sphere	$SA = 4\pi r^2$

PART I

Answer all 28 questions in this part. Each correct answer will receive 2 credits. No partial credit will be allowed. For each question, write in the space provided the number preceding the word or expression that best completes the statement or answers the question. [56 credits]

1 In the diagram below, $\triangle ABC \cong \triangle XYZ$.

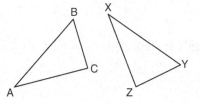

Which two statements identify corresponding congruent parts for these triangles?

(1) $\overline{AB} \cong \overline{XY}$ and $\angle C \cong \angle Y$

(2) $\overline{AB} \cong \overline{YZ}$ and $\angle C \cong \angle X$

(3) $\overline{BC} \cong \overline{XY}$ and $\angle A \cong \angle Y$

(4) $\overline{BC} \cong \overline{YZ}$ and $\angle A \cong \angle X$

1 __4__

2 A support beam between the floor and ceiling of a house forms a 90° angle with the floor. The builder wants to make sure that the floor and ceiling are parallel. Which angle should the support beam form with the ceiling?

(1) 45°   (3) 90°

(2) 60°   (4) 180°

2 __3__

3 In the diagram below, the vertices of $\triangle DEF$ are the midpoints of the sides of equilateral triangle ABC, and the perimeter of $\triangle ABC$ is 36 cm.

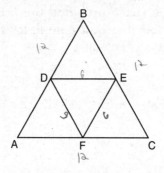

What is the length, in centimeters, of $\overline{EF}$?

(1) 6   (3) 18

(2) 12   (4) 4   3 _1_

4 What is the solution of the following system of equations?

$$y = (x + 3)^2 - 4$$
$$y = 2x + 5$$

(1) (0,–4)   (3) (–4,–3) and (0,5)

(2) (–4,0)   (4) (–3,–4) and (5,0)   4 _3_

$(x+7)(x+3)$

$(x+3)^2 - 4 = 2x+5$

$x^2 + 6x + 9 = 2x + 5$
$ -2x$

$x^2 + 4x + 9 = 5$
$ -5$

$x^2 + 4x + 4 = 0$

$(x+2)(x+2)$

$(x=-2)(x=-2)$

5 One step in a construction uses the endpoints of $\overline{AB}$ to create arcs with the same radii. The arcs intersect above and below the segment. What is the relationship of $\overline{AB}$ and the line connecting the points of intersection of these arcs?

  (1) collinear
  (2) congruent
  (3) parallel
  (4) perpendicular   5 __4__

6 If $\triangle ABC \sim \triangle ZXY$, $m\angle A = 50$, and $m\angle C = 30$, what is $m\angle X$?

  (1) 30   (3) 80
  (2) 50   (4) 100   6 __1__

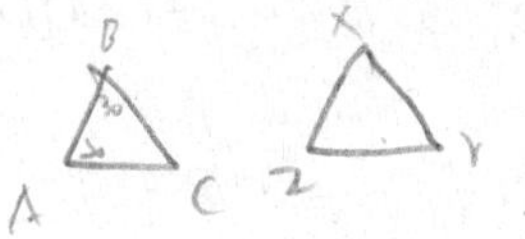

7 In the diagram below of $\triangle AGE$ and $\triangle OLD$, $\angle GAE \cong \angle LOD$, and $\overline{AE} \cong \overline{OD}$.

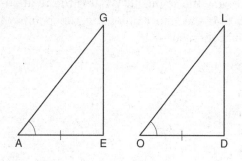

To prove that $\triangle AGE$ and $\triangle OLD$ are congruent by SAS, what other information is needed?

(1) $\overline{GE} \cong \overline{LD}$ (3) $\angle AGE \cong \angle OLD$

(2) $\overline{AG} \cong \overline{OL}$ (4) $\angle AEG \cong \angle ODL$ 7 __2__

8 Point A is not contained in plane $\mathcal{B}$. How many lines can be drawn through point A that will be perpendicular to plane $\mathcal{B}$?

(1) one   (3) zero

(2) two   (4) infinite   ✗8 __3__

9 The equation of a circle is $x^2 + (y - 7)^2 = 16$. What are the center and radius of the circle?

(1) center = (0,7); radius = 4

(2) center = (0,7); radius = 16

(3) center = (0,−7); radius = 4

(4) center = (0,−7); radius = 16   9 __1__

10 What is an equation of the line that passes through the point (7,3) and is parallel to the line $4x + 2y = 10$?

(1) $y = \frac{1}{2}x - \frac{1}{2}$

(3) $y = 2x - 11$

(2) $y = -\frac{1}{2}x + \frac{13}{2}$

(4) $y = -2x + 17$

10 ____

11 In $\triangle ABC$, $AB = 7$, $BC = 8$, and $AC = 9$. Which list has the angles of $\triangle ABC$ in order from smallest to largest?

(1) $\angle A, \angle B, \angle C$

(3) $\angle C, \angle B, \angle A$

(2) $\angle B, \angle A, \angle C$

(4) $\angle C, \angle A, \angle B$

11 __4__

12 Tangents $\overline{PA}$ and $\overline{PB}$ are drawn to circle O from an external point, P, and radii $\overline{OA}$ and $\overline{OB}$ are drawn. If m$\angle APB = 40$, what is the measure of $\angle AOB$?

(1) 140°

(3) 70°

(2) 100°

(4) 50°

12 __3__

13 What is the length of the line segment with endpoints $A(-6,4)$ and $B(2,-5)$?

(1) $\sqrt{13}$

(3) $\sqrt{72}$

(2) $\sqrt{17}$

(4) $\sqrt{145}$

13 __4__

14 The lines represented by the equations $y + \frac{1}{2}x = 4$

and $3x + 6y = 12$ are

$\frac{6-1}{6} = \frac{-3x+12}{6}$ $\frac{}{6}$ $y = -\frac{1}{2}x + 4$

(1) the same line

(2) parallel

(3) perpendicular $y = -\frac{1}{2}x + 2$

(4) neither parallel nor perpendicular

14 ___2___

15 A transformation of a polygon that always preserves both length and orientation is

(1) dilation   (3) line reflection

(2) translation   (4) glide reflection   15 ___2___

16 In which polygon does the sum of the measures of the interior angles equal the sum of the measures of the exterior angles?

(1) triangle   (3) octagon

(2) hexagon   (4) quadrilateral   16 ___4___

17 In the diagram below of circle O, chords $\overline{AB}$ and $\overline{CD}$ intersect at E.

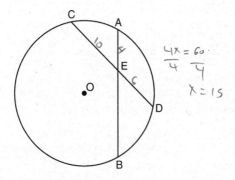

If $CE = 10$, $ED = 6$, and $AE = 4$, what is the length of $\overline{EB}$?

(1) 15   (3) 6.7

(2) 12   (4) 2.4   17 __1__

18 In the diagram below of $\triangle ABC$, medians $\overline{AD}$, $\overline{BE}$, and $\overline{CF}$ intersect at G.

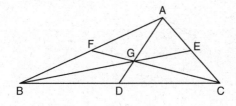

If $CF = 24$, what is the length of $\overline{FG}$?

(1) 8   (3) 12

(2) 10   (4) 16   18 __1__

19 If a line segment has endpoints $A(3x + 5, 3y)$ and $B(x - 1, - y)$, what are the coordinates of the midpoint of $\overline{AB}$?

(1) $(x + 3, 2y)$ (3) $(2x + 3, y)$

(2) $(2x + 2, y)$ (4) $(4x + 4, 2y)$ 19 ___

20 If the surface area of a sphere is represented by 144π, what is the volume in terms of π?

(1) 36π (3) 216π

(2) 48π (4) 288π 20 ___

21 Which transformation of the line $x = 3$ results in an image that is perpendicular to the given line?

(1) $r_{x\text{-axis}}$ (3) $r_{y = x}$

(2) $r_{y\text{-axis}}$ (4) $r_{y = 1}$ 21 ___

22 In the diagram below of regular pentagon $ABCDE$, $\overline{EB}$ is drawn.

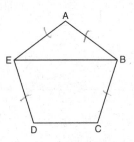

What is the measure of $\angle AEB$?

(1) $36°$ (3) $72°$

(2) $54°$ (4) $108°$ 22 ___

23 $\triangle ABC$ is similar to $\triangle DEF$. The ratio of the length of $\overline{AB}$ to the length of $\overline{DE}$ is 3:1. Which ratio is also equal to 3:1?

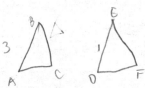

(1) $\dfrac{m\angle A}{m\angle D}$

(2) $\dfrac{m\angle B}{m\angle F}$

(3) $\dfrac{\text{area of } \triangle ABC}{\text{area of } \triangle DEF}$

(4) $\dfrac{\text{perimeter of } \triangle ABC}{\text{perimeter of } \triangle DEF}$

23 __4__

24 What is the slope of a line perpendicular to the line whose equation is $2y = -6x + 8$?

(1) -3 (3) $\dfrac{1}{3}$

(2) $\dfrac{1}{6}$ (4) -6

24 __3__

$$\frac{2y = -6x + 8}{2} \quad \frac{}{2} \quad \frac{}{2}$$

$$y = -3x + 8$$

$$y = \frac{1}{3}x + 8$$

25 In the diagram below of circle C, $m\overset{\frown}{QT} = 140$ and $m\angle P = 40$.

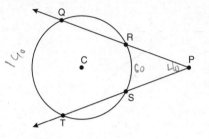

What is $m\overset{\frown}{RS}$?

(1) 50   (3) 90

(2) 60   (4) 100   25 __2__

26 Which statement is logically equivalent to "If it is warm, then I go swimming"?

(1) If I go swimming, then it is warm.

(2) If it is warm, then I do not go swimming.

(3) If I do not go swimming, then it is not warm.

(4) If it is not warm, then I do not go swimming.   ✗ 26 __4__

27 In the diagram below of $\triangle ACT$, $\overleftrightarrow{BE} \parallel \overline{AT}$.

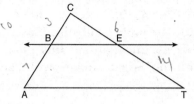

If $CB = 3$, $CA = 10$, and $CE = 6$, what is the length of $\overline{ET}$?

(1) 5   (3) 20
(2) 14   (4) 26   27 __2__

28 Which geometric principle is used in the construction shown below?

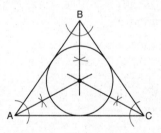

(1) The intersection of the angle bisectors of a triangle is the center of the inscribed circle.
(2) The intersection of the angle bisectors of a triangle is the center of the circumscribed circle.
(3) The intersection of the perpendicular bisectors of the sides of a triangle is the center of the inscribed circle.
(4) The intersection of the perpendicular bisectors of the sides of a triangle is the center of the circumscribed circle.   ✗28 __2__

PART II

Answer all 6 questions in this part. Each correct answer will receive 2 credits. Clearly indicate the necessary steps, including appropriate formula substitutions, diagrams, graphs, charts, etc. For all questions in this part, a correct numerical answer with no work shown will receive only 1 credit.   [12 credits]

29 The diagram below shows isosceles trapezoid $ABCD$ with $\overline{AB} \parallel \overline{DC}$ and $\overline{AD} \cong \overline{BC}$. If $m\angle BAD = 2x$ and $m\angle BCD = 3x + 5$, find $m\angle BAD$.

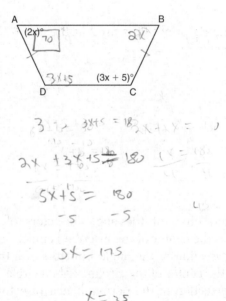

30 A right circular cone has a base with a radius of 15 cm, a vertical height of 20 cm, and a slant height of 25 cm. Find, in terms of π, the number of square centimeters in the lateral area of the cone.

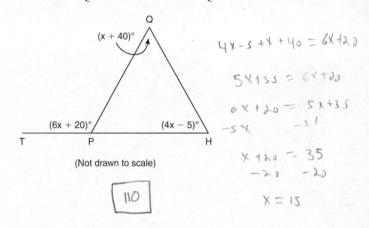

$\pi(15)(25)$

375π

31 In the diagram below of $\triangle HQP$, side $\overrightarrow{HP}$ is extended through P to T, $m\angle QPT = 6x + 20$, $m\angle HQP = x + 40$, and $m\angle PHQ = 4x - 5$. Find $m\angle QPT$.

(Not drawn to scale)

110

$4x - 5 + x + 40 = 6x + 20$

$5x + 35 = 6x + 20$

$6x + 20 = 5x + 35$

$-5x \qquad -5x$

$x + 20 = 35$

$-20 \qquad -20$

$x = 15$

32 On the line segment below, use a compass and straightedge to construct equilateral triangle *ABC*. [Leave all construction marks.]

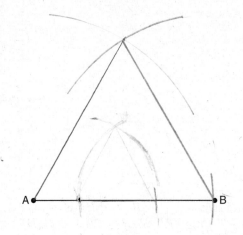

33 In the diagram below, car *A* is parked 7 miles from
   car *B*. Sketch the points that are 4 miles from car
   A and sketch the points that are 4 miles from
   car *B*. Label with an **X** all points that satisfy both
   conditions.

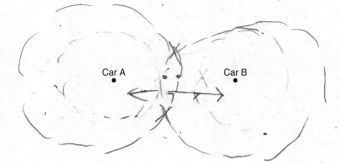

34 Write an equation for circle O shown on the graph below.

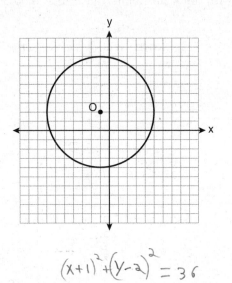

$$(x+1)^2 + (y-2)^2 = 36$$

PART III

Answer all 3 questions in this part. Each correct answer will receive 4 credits. Clearly indicate the necessary steps, including appropriate formula substitutions, diagrams, graphs, charts, etc. For all questions in this part, a correct numerical answer with no work shown will receive only 1 credit. [12 credits]

35 In the diagram below of quadrilateral $ABCD$ with diagonal $\overline{BD}$, m$\angle A$ = 93, m$\angle ADB$ = 43, m$\angle C$ = $3x + 5$, m$\angle BDC = x + 19$, and m$\angle DBC = 2x + 6$. Determine if $\overline{AB}$ is parallel to $\overline{DC}$. Explain your reasoning.

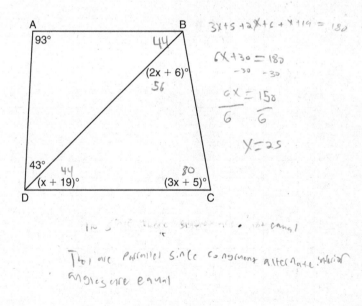

36 The coordinates of the vertices of $\triangle ABC$ are $A(1,3)$, $B(-2,2)$, and $C(0,-2)$. On the grid below, graph and label $\triangle A''B''C''$, the result of the composite transformation $D_2 \circ T_{3,-2}$. State the coordinates of A'', B'', and C''.

$A' = 4,1$
$B' = 1,0$
$C' = 3,-4$

$A'' = 8,2$
$B'' = 2,0$
$C'' = 6,-8$

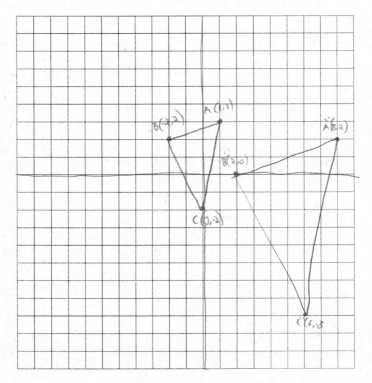

37 In the diagram below, $\triangle RST$ is a 3-4-5 right triangle. The altitude, h, to the hypotenuse has been drawn. Determine the length of h.

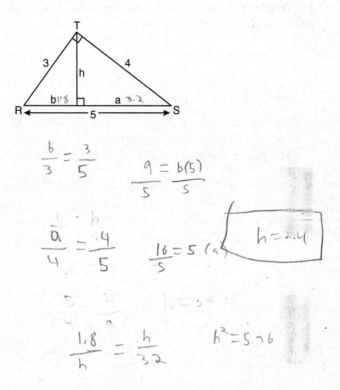

$$\frac{b}{3} = \frac{3}{5} \qquad \frac{9}{5} = \frac{b(5)}{5}$$

$$\frac{a}{4} = \frac{4}{5} \qquad \frac{16}{5} = 5 \, (a$$

$$\boxed{h = 2.4}$$

$$16 = 5a$$

$$\frac{1.8}{h} = \frac{h}{3.2} \qquad h^2 = 5.76$$

PART IV

Answer the question in this part. A correct answer will receive 6 credits. Clearly indicate the necessary steps, including appropriate formula substitutions, diagrams, graphs, charts, etc. A correct numerical answer with no work shown will receive only 1 credit. [6 credits]

38 Given: Quadrilateral $ABCD$ has vertices $A(-5,6)$, $B(6,6)$, $C(8,-3)$, and $D(-3,-3)$.

Prove: Quadrilateral $ABCD$ is a parallelogram but is neither a rhombus nor a rectangle.
[The use of the grid below is optional.]

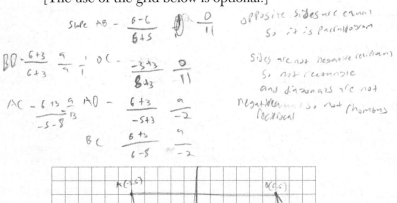

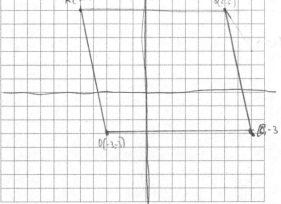

Answers
August 2010
Geometry

Answer Key

PART I

1. (4)	**8.** (1)	**15.** (2)	**22.** (1)
2. (3)	**9.** (1)	**16.** (4)	**23.** (4)
3. (1)	**10.** (4)	**17.** (1)	**24.** (3)
4. (3)	**11.** (4)	**18.** (1)	**25.** (2)
5. (4)	**12.** (1)	**19.** (2)	**26.** (3)
6. (4)	**13.** (4)	**20.** (4)	**27.** (2)
7. (2)	**14.** (2)	**21.** (3)	**28.** (1)

PART II

29. 70

30. 375π

31. 110

32. See the *Answers Explained* section.

33. Two points of intersection. See the *Answers Explained* section.

34. $(x + 1)^2 + (y - 2)^2 = 36$

PART III

35. Parallel. See the *Answers Explained* section.

36. $A''(8,2)$, $B''(2,0)$, and $C''(6,-8)$

37. 2.4

PART IV

38. See the *Answers Explained* section.

In **PARTS II–IV** you are required to show how you arrived at your answers. For sample methods of solutions, see the *Answers Explained* section.

Answers Explained

PART I

1. It is given in the accompanying diagram that $\triangle ABC \cong \triangle XYZ$.

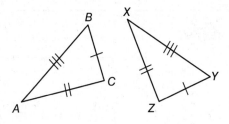

You are asked to identify from the set of answer choices two pairs of congruent parts. When two triangles are congruent, congruent angles are located opposite congruent sides.

Congruent Sides	Congruent Angles
$\overline{BC} \cong \overline{YZ}$	$\angle A \cong \angle X$
$\overline{AC} \cong \overline{XZ}$	$\angle B \cong \angle Y$
$\overline{AB} \cong \overline{XY}$	$\angle C \cong \angle Z$

Compare the pairs of congruent parts listed in each choice with the preceding table.

- Choice (1): $\overline{AB} \cong \overline{XY}$ and $\angle C \cong \angle Y$.   ✗
- Choice (2): $\overline{AB} \cong \overline{YZ}$ and $\angle C \cong \angle X$.   ✗
- Choice (3): $\overline{BC} \cong \overline{XY}$ and $\angle A \cong \angle Y$.   ✗
- Choice (4): $\overline{BC} \cong \overline{YZ}$ and $\angle A \cong \angle X$.   ✔

The correct choice is **(4)**.

2. It is given that a support beam between the floor and ceiling of a house forms a 90° angle with the floor. If two planes (floor and ceiling) are perpendicular to the same line (the support beam), they are parallel. If the builder wants to make sure that the floor and ceiling are parallel, the support beam should also form a 90° angle with the ceiling.

The correct choice is **(3)**.

3. In the accompanying diagram of equilateral $\triangle ABC$, the vertices of $\triangle DEF$ are the midpoints of the sides of $\triangle ABC$.

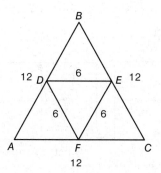

Since it is also given that the perimeter of $\triangle ABC$ is 36 cm, each side of $\triangle ABC$ measures 12 cm.

Because the segment joining the midpoints of two sides of a triangle is one-half the length of the third side of the triangle, each side of $\triangle DEF$ measures $\frac{1}{2} \times 12$ cm = 6 cm. Hence, the length of $\overline{EF}$ is 6 cm.

The correct choice is **(1)**.

4. You are asked to determine the solution to the following system of equations:

$$y = (x + 3)^2 - 4$$
$$y = 2x + 5$$

Method 1: Solve graphically using your graphing calculator. Select a friendly window in which the graphs fit, such as $X_{min} = -9.4$, $X_{max} = 9.4$, $Y_{min} = -6.2$, and $Y_{max} = 6.2$. Then graph $Y_1 = (x + 3)^2 - 4$ and $Y_2 = 2x + 5$ on the same set of axes. Use the Trace or Intersect features of your calculator to find that the graphs intersect at $(-4, -3)$ and $(0, 5)$.

The correct choice is **(3)**.

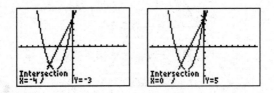

Method 2: Solve algebraically. Eliminate y by replacing it in the first equation with $2x + 5$:

$$2x + 5 = (x + 3)^2 - 4$$

Square the binomial. Then simplify by collecting all of the nonzero terms on the same side of the equation:

$$2x + 5 = x^2 + 6x + 9 - 4$$
$$2x + 5 = x^2 + 6x + 5$$
$$0 = x^2 + (6x - 2x) + (5 - 5)$$
$$0 = x^2 + 4x$$

Solve the quadratic equation by factoring:

$$x(x + 4) = 0$$
$$x = 0 \quad \text{or} \quad x + 4 = 0$$
$$x = -4$$

Since answer choice (3) is the only choice that includes points with x-coordinates of 0 and -4, it is the correct choice.

The correct choice is **(3)**.

5. It is given that one step in a construction uses the endpoints of $\overline{AB}$ to create arcs with the same radii such that the arcs intersect above and below the segment, as illustrated in the accompanying diagram.

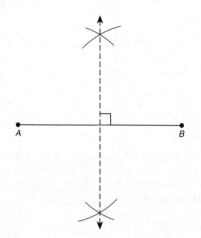

The points at which the two pairs of arcs intersect determine a line that is perpendicular to $\overline{AB}$.

The correct choice is (**4**).

6. It is given that $\triangle ABC \sim \triangle ZXY$. Since the sum of the measures of the three angles of a triangle is 180°:

$$m\angle A + m\angle B + m\angle C = 180$$
$$50 + m\angle B + m\angle 30 = 180$$
$$m\angle B = 180 - 80$$
$$= 100$$

Angles B and X are corresponding angles since both letters occur in the second position in the similarity relation $\triangle ABC \sim \triangle ZXY$. Because corresponding angles of similar triangles are congruent, $m\angle X = m\angle B = 100$.

The correct choice is (**4**).

7. In the accompanying diagram, it is given that $\angle GAE \cong \angle LOD$ and $\overline{AE} \cong \overline{OD}$.

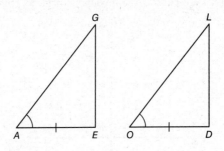

In order to prove that $\triangle AGE$ and $\triangle OLD$ are congruent by SAS (side-angle-side), you must know that the two sides that form $\angle GAE$ are congruent to the corresponding sides that form $\angle LOD$. In other words, in addition to $\overline{AE} \cong \overline{OD}$, we would also need to know that $\overline{AG} \cong \overline{OL}$.

The correct choice is (**2**).

8. If point A is not contained in plane $\mathcal{B}$, exactly one line can be drawn through A and perpendicular to plane $\mathcal{B}$, as illustrated in the accompanying diagram.

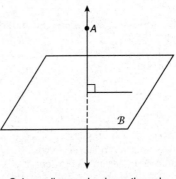

Only one line can be drawn through
A and perpendicular to plane $\mathcal{B}$.

The correct choice is (**1**).

9. The general equation of a circle has the form $(x - h)^2 + (y - k)^2 = r^2$, where (h,k) is the center of the circle and r is the radius. The given equation, $x^2 + (y - 7)^2 = 16$, can be rewritten as $(x - 0)^2 + (y - 7)^2 = 4^2$, where $(h,k) = (0,7)$ and $r = 4$. Hence, the center of the given circle is $(0,7)$ and its radius is 4.

The correct choice is (**1**).

10. You are required to write an equation of the line that passes through the point $(7,3)$ and is parallel to the line $4x + 2y = 10$.

- If $4x + 2y = 10$, then $2y = -4x + 10$ and $y = \dfrac{-4}{2}x + \dfrac{10}{2}$ or $y = -2x + 5$. In general, a line whose equation is in the slope-intercept form $y = mx + b$ has a slope of m and a y-intercept of b. The slope of the line $y = -2x + 5$ is therefore -2.

- Since parallel lines have equal slopes, the slope of the required line is also -2. The required line has an equation of the form $y = -2x + b$. Find b by substituting the coordinates of $(7,3)$ into the equation $y = -2x + b$:

$$3 = -2(7) + b$$
$$3 = -14 + b$$
$$17 = b$$

- Since $m = -2$ and $b = 17$, an equation of the required line is $y = -2x + 17$.

The correct choice is (**4**).

11. In a triangle, the larger angle lies opposite the longer side. It is given that in $\triangle ABC$, $AB = 7$, $BC = 8$, and $AC = 9$, as shown in the accompanying diagram.

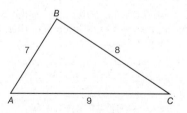

Since $AB < BC < AC$, the angles opposite these sides have the same size relationship, $m\angle C < m\angle A < m\angle B$. In order from smallest to largest, the three angles of the triangle are $\angle C$, $\angle A$, $\angle B$.

The correct choice is (**4**).

12. It is given that tangents $\overline{PA}$ and $\overline{PB}$ are drawn to circle O from the same external point P, m$\angle APB = 40$, and radii $\overline{OA}$ and $\overline{OB}$ are drawn, as shown in the accompanying diagram.

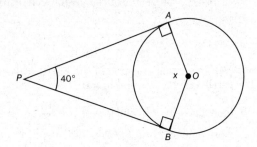

Since a radius drawn to a tangent at the point of contact is perpendicular to the tangent, angles A and B are right angles.

The sum of the measures of the four angles of quadrilateral $PAOB$ is 360°. If x represents the measure of $\angle AOB$:

$$40 + 90 + x + 90 = 360$$
$$x + 220 = 360$$
$$x = 360 - 220$$
$$= 140$$

The correct choice is (**1**).

13. To find the length, d, of the line segment with endpoints are $A(-6,4)$ and $B(2,-5)$, use the distance formula:

$$d = \sqrt{\left(\Delta x\right)^2 + \left(\Delta y\right)^2}$$

In this formula, Δx represents the difference in the x-coordinates of the two given points and Δy represents the difference in their y-coordinates, with the subtractions performed in the same order:

$$AB = \sqrt{\left(2 - (-6)\right)^2 + \left((-5) - 4\right)^2}$$
$$= \sqrt{\left(2 + 6\right)^2 + (-9)^2}$$
$$= \sqrt{64 + 81}$$
$$= \sqrt{145}$$

The correct choice is (**4**).

14. The equations of the given lines are $y + \frac{1}{2}x = 4$ and $3x + 6y = 12$. To determine the relationship between the two lines, write each equation in $y = mx + b$ slope-intercept form where m is the slope of the line and b is its y-intercept.

- If $y + \frac{1}{2}x = 4$, then $y = -\frac{1}{2}x + 4$. Hence, $m = -\frac{1}{2}$ and $b = 4$.

- If $3x + 6y = 12$, then $6y = -3x + 12$ and $\frac{6y}{6} = \frac{-3x}{6} + \frac{12}{6}$ so $y = -\frac{1}{2}x + 2$. Hence, $m = -\frac{1}{2}$ and $b = 2$.

- Since the two lines have the same slope but different y-intercepts, they are parallel lines.

The correct choice is **(2)**.

15. You are required to identify the type of transformation of a polygon that always preserves both length and orientation.

- Since a reflection always reverses orientation, you can eliminate line reflection in choice (3) and glide reflection in choice (4).

- Under a dilation, the lengths of the sides of the polygon are multiplied by some fixed scale factor. Hence, you can eliminate dilation in choice (1).

- A translation, choice (2), preserves both length and orientation.

The correct choice is **(2)**.

16. You are asked to identify the polygon for which the sum of the measures of the interior angles equals the sum of the measures of the exterior angles. Regardless of how many sides a polygon has, the sum of the measures of its exterior angles is $360°$. The sum of the measures of the interior angles of a quadrilateral is $360°$.

The correct choice is **(4)**.

17. In the accompanying diagram of circle O, chords $\overline{AB}$ and $\overline{CD}$ intersect at E such that $CE = 10$, $ED = 6$, and $AE = 4$. You are asked to find the length of $\overline{EB}$.

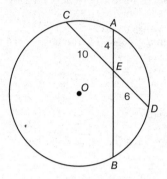

If two chords intersect inside a circle, the product of the lengths of the segments of one chord is equal to the product of the lengths of the segments of the other chord:

$$AE \times EB = CE \times ED$$
$$4 \times EB = 10 \times 6$$
$$EB = \frac{60}{4}$$
$$= 15$$

The correct choice is **(1)**.

18. In the accompanying diagram of $\triangle ABC$, medians $\overline{AD}$, $\overline{BE}$, and $\overline{CF}$ intersect at G. Given that $CF = 24$, you are asked to find the length of $\overline{FG}$.

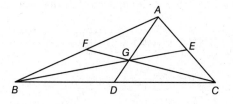

The point at which the three medians of a triangle intersect divides each median into two segments whose lengths are in the ratio 2:1, with the longest segment located nearest the vertex.

If x represents the length of $\overline{FG}$, then $2x$ represents the length of $\overline{CG}$. Since it is given that $CF = 24$:

$$x + 2x = 24$$
$$3x = 24$$
$$x = \frac{24}{3}$$
$$= 8$$

The length of $\overline{FG}$ is 8.

The correct choice is (**1**).

19. It is given that a line segment has endpoints $A(3x + 5, 3y)$ and $B(x-1,-y)$. You are asked to find the coordinates of the midpoint of $\overline{AB}$. The x and y coordinates of the midpoint $M(\overline{x}, \overline{y})$ of a line segment are the averages of the corresponding coordinates of the endpoints of the line segment:

$$M(\overline{x}, \overline{y}) = M\left(\frac{(3x+5)+(x-1)}{2}, \frac{3y+(-y)}{2}\right)$$

$$= M\left(\frac{4x+4}{2}, \frac{2y}{2}\right)$$

$$= M(2x+2, y)$$

The correct choice is **(2)**.

20. The reference sheet provided in the test booklet gives the formula for the surface area (SA) of a sphere as $SA = 4\pi r^2$ and the formula for the volume (V) of a sphere as $V = \frac{4}{3}\pi r^3$, where r is the radius of the sphere.

If the surface area of a sphere is 144π:

$$4\pi r^2 = 144\pi$$

$$r^2 = \frac{144\pi}{4\pi} = 36$$

$$r = \sqrt{36}$$

$$= 6$$

To find the volume of this sphere, evaluate $V = \frac{4}{3}\pi r^3$ using $r = 6$:

$$V = \frac{4}{3}\pi(6^3)$$

$$= \frac{4}{\cancel{3}}\pi(\cancel{216}^{72})$$

$$= 288\pi$$

The correct choice is **(4)**.

21. You are asked to determine from the set of answer choices which transformation of the line $x = 3$ results in an image that is perpendicular to the given line. The line $x = 3$ is a vertical line, parallel to the y-axis and 3 units to the right of it. Examine each of the answer choices in turn.

- Choice (1): Reflecting the given line in the x-axis flips the line over the x-axis so that the image coincides with the given line. ✗

- Choice (2): Reflecting the given line in the y-axis flips the line over the y-axis. The image that results is the line $x = -3$, which is parallel to the given line. ✗

- Choice (3): The image of any point (x,y) reflected in the line $y = x$ is the point (y,x). The image of the reflection of $x = 3$ in the line $y = x$ is the line $y = 3$, which is perpendicular to the given line. See the accompanying figure. ✔

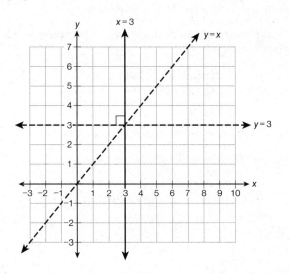

- Choice (4): The given line is two units to the right of the reflecting line $x = 1$. Reflecting the given line in the line $x = 1$ produces an image that is parallel to the given line and 2 units to the left of $x = 1$. ✗

The correct choice is **(3)**.

22. In the accompanying diagram of regular pentagon *ABCDE*, $\overline{EB}$ is drawn. You are asked to find the measure of $\angle AEB$.

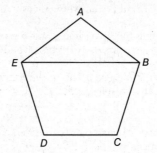

Since a regular polygon is equilateral, $\overline{AE} \cong \overline{AB}$. As a result, $\angle AEB \cong \angle ABE$. If *x* represents m$\angle AEB$, then m$\angle ABE$ = m$\angle AEB$ = *x*.

The sum, *S*, of the measures of the angles of a regular polygon is $S = (n - 2) \times 180°$. For a pentagon, $n = 5$:

$$S = (5 - 2) \times 180$$
$$= 3 \times 180$$
$$= 540$$

Since a regular polygon is equiangular, each angle measures $\dfrac{540°}{5} = 108°$.

Hence, m$\angle A = 108°$. In isosceles $\triangle EAB$:

$$x + 108 + x = 180$$
$$2x = 180 - 108$$
$$\frac{2x}{2} = \frac{72}{2}$$
$$x = 36$$

The correct choice is (**1**).

23. It is given that $\triangle ABC \sim \triangle DEF$ and that the ratio of the length of $\overline{AB}$ to the length of $\overline{DE}$ is 3:1. You are asked to determine from the set of answer choices the ratio that is also equal to 3:1.

- Choice (1): $\dfrac{m\angle A}{m\angle D}$. Since corresponding angles of similar triangles are

  congruent, you can eliminate this choice as $\dfrac{m\angle A}{m\angle D} = 1$. ✗

- Choice (2): $\dfrac{m\angle B}{m\angle F}$. Angles B and F are not corresponding angles. Insufficient information is given to determine their ratio. ✗

- Choice (3): $\dfrac{\text{area of } \triangle ABC}{\text{area of } \triangle DEF}$. The areas of two similar triangles have the

  same ratio as the square of the ratio of the lengths of a pair of corresponding sides. Hence, $\dfrac{\text{area of } \triangle ABC}{\text{area of } \triangle DEF} = \left(\dfrac{3}{1}\right)^2 = \dfrac{9}{1}$ or 9:1. ✗

- Choice (4): $\dfrac{\text{perimeter of } \triangle ABC}{\text{perimeter of } \triangle DEF}$. The perimeters of two similar trian-

  gles have the same ratio as the ratio of the lengths of any pair of corre-

  sponding sides. Hence, $\dfrac{\text{perimeter of } \triangle ABC}{\text{perimeter of } \triangle DEF} = \dfrac{3}{1}$ or 3:1. ✔

The correct choice is **(4)**.

24. Perpendicular lines have slopes that are negative reciprocals. To find the slope of a line perpendicular to the line $2y = -6x + 8$, find the negative reciprocal of the slope of the given line.

In general, a line whose equation is in the $y = mx + b$ slope-intercept form has a slope of m and a y-intercept of b. If $2y = -6x + 8$, then $y = \dfrac{-6x}{2} + \dfrac{8}{2} = -3x + 4$. So the slope of the given line is -3.

Since the negative reciprocal of -3 is $\dfrac{1}{3}$, a line perpendicular to the given line has a slope of $\dfrac{1}{3}$.

The correct choice is **(3)**.

25. In the accompanying diagram of circle C, $m\widehat{QT} = 140$ and $m\angle P = 40$. You need to find $m\widehat{RS}$.

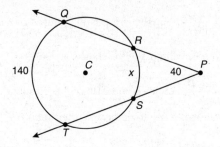

The measure of an angle formed by two secants intersecting in the exterior of a circle is equal to one-half the difference of the measures of the intercepted arcs. If $x = m\widehat{RS}$:

$$m\angle P = \frac{1}{2}\left(m\widehat{QT} - m\widehat{RS}\right)$$

$$40 = \frac{1}{2}(140 - x)$$

$$2(40) = 2\left(\frac{1}{2}(140 - x)\right)$$

$$80 = 140 - x$$

$$x = 140 - 80$$

$$= 60$$

The correct choice is **(2)**.

26. A conditional statement and its contrapositive are logically equivalent statements as they always have the same truth value. To form the contrapositive, interchange and then negate the "if" and "then" parts of the original conditional:

Original conditional: "If <u>it is warm</u>, then <u>I go swimming</u>"

Contrapositive: "If <u>I do not go swimming</u>, then <u>it is not warm</u>"

The correct choice is **(3)**.

27. It is given that in the accompanying diagram of $\triangle ACT$, $\overline{BE} \parallel \overline{AT}$, $CB = 3$, $CA = 10$, and $CE = 6$. You need to find the length of $\overline{ET}$.

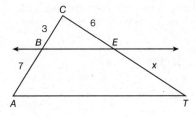

$$BA = CA - CB$$
$$= 10 - 3$$
$$= 7$$

If a line is parallel to one side of a triangle and intersects the other two sides, it divides those intersected sides proportionally. If $x = ET$:

$$\frac{CB}{BA} = \frac{CE}{ET}$$

$$\frac{3}{7} = \frac{6}{x}$$

$$\frac{3x}{3} = \frac{42}{3}$$

$$x = 14$$

The correct choice is **(2)**.

28. You are asked to determine from the set of answer choices the geometric principle used in the accompanying construction.

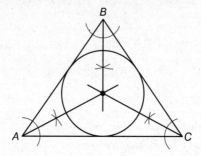

The construction shows that the bisectors of the three angles of a triangle intersect at a single point, called the incenter, which is the center of the inscribed circle.

The correct choice is **(1)**.

PART II

29. It is given that in the accompanying diagram of isosceles trapezoid $ABCD$, $\overline{AB} \parallel \overline{DC}$, $\overline{AD} \cong \overline{BC}$, $m\angle BAD = 2x$, and $m\angle BCD = 3x + 5$. You must find $m\angle BAD$.

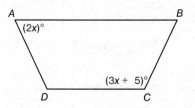

Since base angles of an isosceles trapezoid are equal in measure, $m\angle B = m\angle BAD = 2x$. Because each pair of upper and lower base angles of a trapezoid are supplementary:

$$m\angle B + m\angle BCD = 180$$
$$(2x) + (3x + 5) = 180$$
$$5x + 5 = 180$$
$$5x = 180 - 5$$
$$\frac{5x}{5} = \frac{175}{5}$$
$$x = 35$$

Hence, $m\angle BAD = 2x = 2(35) = \mathbf{70}$.

30. It is given that a right circular cone has a base with a radius of 15 cm, a vertical height of 20 cm, and a slant height of 25 cm. You are required to find, in terms of π, the number of square centimeters in the lateral area of the cone.

As shown on the accompanying reference sheet in your test booklet, the lateral area (L) of a right circular cone is given by the formula $L = \pi r \ell$, where ℓ is the slant height and r is the radius of the base of the cone. Evaluate the formula using $r = 15$ cm and $\ell = 25$ cm:

$$L = \pi(15 \text{ cm})(25 \text{ cm})$$
$$= 375\pi \text{ cm}^2$$

The lateral area of the cone is **375π cm²**.

31. In the accompanying diagram of $\triangle HQP$, m$\angle QPT = 6x + 20$, m$\angle HQP = x + 40$ and m$\angle PHQ = 4x - 5$.

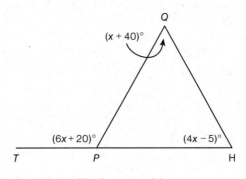

(Not drawn to scale)

To find m$\angle QPT$, use the theorem that states the measure of an exterior angle of a triangle is equal to the sum of the measures of the two nonadjacent (remote) interior angles:

$$\text{m}\angle QPT = \text{m}\angle PHQ + \text{m}\angle HQP$$
$$6x + 20 = (4x - 5) + (x + 40)$$
$$6x + 20 = 5x + 35$$
$$6x - 5x = 35 - 20$$
$$x = 15$$

Thus, m$\angle QPT = 6x + 20 = 6(15) + 20 = \mathbf{110}$.

32. You are required to construct equilateral triangle *ABC* on the given ne segment.

- Step 1: Set the radius length of your compass equal to the length of $\overline{AB}$, the given line segment, by placing the fixed point of the compass on *A* and the pencil point on *B*. Using this compass setting and keeping the point of the compass fixed on *A*, draw an arc.

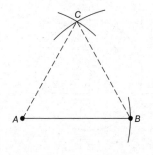

- Step 2: Shift the compass point to *B*. Using the same radius length, draw an arc that intersects the first arc. Label the point of intersection, *C*.

- Step 3: Draw $\overline{AC}$ and $\overline{BC}$, thereby forming **equilateral △*ABC*.**

33. It is given that in the accompanying diagram, car *A* is parked 7 miles om car *B*. You are asked to sketch the points that are 4 miles from car *A* and sketch the points that are 4 miles from car *B*. You also need to label with an all points that satisfy both conditions.

The locus of points that are a given distance from a point is a circle with 1e given point as the center and the distance as the radius length. Using a ompass, construct circles with the same radii having centers at points *A* and , as shown in the accompanying diagram.

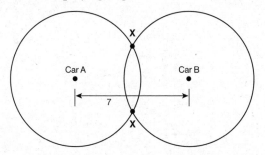

Since 7 < 4 + 4, the two circles intersect at two different points. Label the vo points of intersection with an **X**.

34. The general form of an equation of a circle with center at (h,k) and radius r is $(x - h)^2 + (y - k)^2 = r^2$. To determine an equation of the circle in the accompanying graph, read the coordinates of the center from the graph and determine the length of the radius.

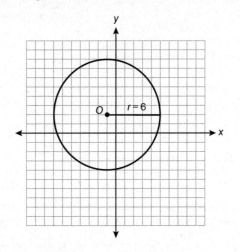

As no scale is indicated, you can assume for this exam that each grid box represents 1 unit. The center of the circle is at $O(-1,2)$. To determine the radius, count the number of unit boxes along a grid line from O to the circumference of the circle. The radius is 6 units. Write an equation of the circle using $(h,k) = (-1,2)$ and $r = 6$:

$$(x - (-1))^2 + (y - 2)^2 = 6^2$$
$$(x + 1)^2 + (y - 2)^2 = 36$$

PART III

35. In the accompanying diagram of quadrilateral $ABCD$, $m\angle A = 93$, $m\angle ADB = 43$, $m\angle C = 3x + 5$, $m\angle BDC = x + 19$, and $m\angle DBC = 2x + 6$.

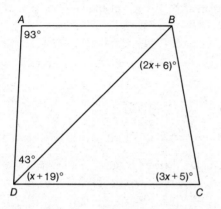

To determine if $\overline{AB}$ is parallel to $\overline{DC}$, find and then compare the measures of alternate interior angles ABD and BDC.

Since the sum of the measures of the angles of a triangle is 180, in $\triangle BAD$:

$$43 + 93 + m\angle ABD = 180$$
$$m\angle ABD = 180 - 136$$
$$= 44$$

In $\triangle BCD$,

$$(2x + 6) + (3x + 5) + (x + 19) = 180$$
$$6x + 30 = 180$$
$$6x = 180 - 30$$
$$\frac{6x}{6} = \frac{150}{6}$$
$$x = 25$$

Thus, $m\angle BDC = x + 19 = 25 + 19 = 44$. Since $m\angle ABD = 44$ and $m\angle BDC = 44$, $\angle ABD \cong \angle BDC$. When parallel line segments are cut by a transversal, like $\overline{BD}$, congruent alternate interior angles are formed. So $\overline{AB}$ is **parallel** to $\overline{DC}$.

36. The coordinates of the vertices of $\triangle ABC$ are $A(1,3)$, $B(-2,2)$, and $C(0,-2)$. You are required to graph and label $\triangle A''B''C''$, the result of the composite transformation $D_2 \circ T_{3,-2}$. To evaluate $D_2 \circ T_{3,-2}$, first translate $\triangle ABC$ and then dilate the resulting image.

Under the translation $T_{3,-2}$, each of the vertices of $\triangle ABC$ are shifted 3 units to the right and 2 units down:

$$A(1,3) \xrightarrow{\;T_{3,-2}\;} (1+3,3+(-2)) = A'(4,1)$$

$$B(-2,2) \xrightarrow{\;T_{3,-2}\;} (-2+3,2+(-2)) = B'(1,0)$$

$$C(0,-2) \xrightarrow{\;T_{3,-2}\;} (0+3,-2+(-2)) = C'(3,-4)$$

Under the dilation D_2, the coordinates of the vertices of $\triangle A'B'C'$ are each multiplied by 2:

$$A'(4,1) \xrightarrow{\;D_2\;} (4\cdot 2,1\cdot 2) = A''(8,2)$$

$$B'(1,0) \xrightarrow{\;D_2\;} (1\cdot 2, 0\cdot 2) = B''(2,0)$$

$$C'(3,-4) \xrightarrow{\;D_2\;} (3\cdot 2,-4\cdot 2) = C''(6,-8)$$

Next, graph $\triangle A''B''C''$ on the grid provided in the test booklet, as shown in the accompanying diagram. Label each of the vertices of the triangle with its coordinates: $A''(8,2)$, $B''(2,0)$, and $C''(6,-8)$.

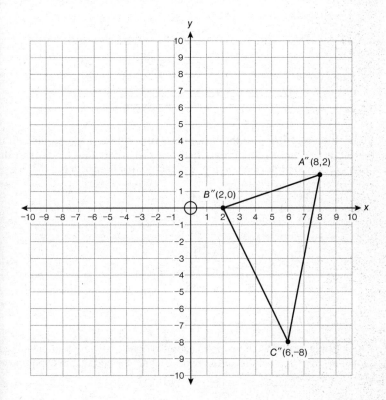

37. In the accompanying diagram, $\triangle RST$ is a 3-4-5 right triangle with the altitude, h, drawn to the hypotenuse. You are asked to find the length of h.

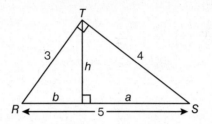

If an altitude is drawn to the hypotenuse of a right triangle, the length of either leg is the mean proportional between the hypotenuse length and the length of the segment formed on the hypotenuse that is adjacent to that leg:

$$\frac{5}{4} = \frac{4}{a}$$

$$5a = 16$$

$$\frac{5a}{5} = \frac{16}{5}$$

$$a = 3.2$$

Since $a = 3.2$, $b = 5 - a = 5 - 3.2 = 1.8$.

The altitude drawn to the hypotenuse of a right triangle is the mean proportional between the lengths of the two segments formed on the hypotenuse:

$$\frac{a}{h} = \frac{h}{b}$$

$$\frac{3.2}{h} = \frac{h}{1.8}$$

$$h^2 = 3.2 \times 1.8$$

$$\sqrt{h^2} = \sqrt{5.76}$$

$$h = 2.4$$

The length of altitude h is **2.4**.

ART IV

38. You are given that Quadrilateral $ABCD$ has vertices $A(-5,6)$, $B(6,6)$, $(8,-3)$, and $D(-3,-3)$. You are required to prove that Quadrilateral $ABCD$ is parallelogram but is neither a rhombus nor a rectangle.

Plan: Although you can complete the proof in a number of different ways, ne of the simplest ways is to work with the diagonals of the figure. First rove that quadrilateral $ABCD$ is a parallelogram by using the midpoint for-ula to show that its diagonals bisect each other. A rhombus has the property nat its diagonals intersect at right angles, and a rectangle has the property nat its diagonals have the same length. Using the slope and distance formu-s, show that the diagonals of parallelogram $ABCD$ do not have these two roperties and, as result, is neither a rhombus nor a rectangle.

Prove $ABCD$ is a parallelogram:

- Midpoint of diagonal $\overline{AC} = \left(\dfrac{-5+8}{2}, \dfrac{6+(-3)}{2} \right) = \left(\dfrac{3}{2}, \dfrac{3}{2} \right)$

- Midpoint of diagonal $\overline{BD} = \left(\dfrac{6+(-3)}{2}, \dfrac{6+(-3)}{2} \right) = \left(\dfrac{3}{2}, \dfrac{3}{2} \right)$

- Since the diagonals of the quadrilateral have the same midpoint, they bisect each other. If the diagonals of a quadrilateral bisect each other, the quadrilateral is a parallelogram. Hence, quadrilateral $ABCD$ is a parallelogram.

Note that you could also prove that $ABCD$ is a parallelogram using one of ne following three methods: You could show that opposite sides have the ame slope and, as a result, are parallel. You could show that opposite sides ave the same length and, as a result, are congruent. You could show that one air of sides is both parallel and congruent.

Prove parallelogram $ABCD$ is a not a rhombus:

- Slope of diagonal $\overline{AC} = \dfrac{\Delta y}{\Delta x} = \dfrac{-3-6}{8-(-5)} = \dfrac{-9}{8+5} = -\dfrac{9}{13}$

- Slope of diagonal $\overline{BD} = \dfrac{\Delta y}{\Delta x} = \dfrac{-3-6}{-3-6} = \dfrac{-9}{-9} = 1$

- Since the slopes of the diagonals are not negative reciprocals, the diagonals do not intersect at right angles. As result, parallelogram $ABCD$ not a rhombus.

Note that you could also prove that parallelogram $ABCD$ is not a rhombus by using the distance formula to show that a pair of adjacent sides do not have the same length.

Prove parallelogram $ABCD$ is a not a rectangle:

$$\text{Length of } \overline{AC} = \sqrt{(\Delta x)^2 + (\Delta y)^2}$$

$$= \sqrt{(8 - (-5))^2 + (-3 - 6)^2}$$

$$= \sqrt{(8 + 5)^2 + (-9)^2}$$

$$= \sqrt{169 + 81}$$

$$= \sqrt{250}$$

$$\text{Length of } \overline{BD} = \sqrt{(\Delta x)^2 + (\Delta y)^2}$$

$$= \sqrt{(-3 - 6)^2 + (-3 - 6)^2}$$

$$= \sqrt{(-9)^2 + (-9)^2}$$

$$= \sqrt{81 + 81}$$

$$= \sqrt{162}$$

Since $\sqrt{250} \neq \sqrt{162}$, $\overline{AC} \not\cong \overline{BD}$. Because its diagonals are not congruent parallelogram $ABCD$ is not a rectangle.

Note that you could also prove that parallelogram $ABCD$ is not a rectangle by using the slope formula to show that a pair of adjacent sides do not have slopes that are negative reciprocals and, as a result, are not perpendicular to each other.

Topic	Question Numbers	Number of Points	Your Points	Your Percentage
1. Logic (negation, conjunction, disjunction, related conditionals, biconditional, logically equivalent statements, logical inference)	26	2		
2. Angle & Line Relationships (vertical angles, midpoint, altitude, median, bisector, supplementary angles, complementary angles, parallel and perpendicular lines)	2, 35	$2 + 4 = 6$		
3. Parallel & Perpendicular Planes	8	2		
4. Angles of a Triangle & Polygon (sum of angles, exterior angle, base angles theorem, angles of a regular polygon)	16, 22, 31	$2 + 2 + 2 = 6$		
5. Triangle Inequalities (side length restrictions, unequal angles opposite unequal sides, exterior angle)	11	2		
6. Trapezoids & Parallelograms (includes properties of rectangle, rhombus, square, and isosceles trapezoid)	29	2		
7. Proofs Involving Congruent Triangles	1, 7	$2 + 2 = 4$		
8. Indirect Proof & Mathematical Reasoning	—	—		
9. Ratio & Proportion (includes similar polygons, proportions, proofs involving similar triangles, comparing linear dimensions, and areas of similar triangles)	6, 23, 27	$2 + 2 + 2 = 6$		
10. Proportions formed by altitude to hypotenuse of right triangle; Pythagorean theorem	37	4		
11. Coordinate Geometry (slopes of parallel and perpendicular lines; slope-intercept and point-slope equations of a line; applications requiring midpoint, distance, or slope formulas; coordinate proofs)	10, 13, 14, 19, 24, 38	$2 + 2 + 2 + 2 + 2 + 6 = 16$		
12. Solving a Linear-Quadratic System of Equations Graphically	4	2		
13. Transformation Geometry (includes isometries, dilation, symmetry, composite transformations, transformations using coordinates)	15, 21, 36	$2 + 2 + 4 = 8$		
14. Locus & Constructions (simple and compound locus, locus using coordinates, compass constructions)	5, 32, 33	$2 + 2 + 2 = 6$		

Topic	Question Numbers	Number of Points	Your Points	Your Percentag
15. Midpoint and Concurrency Theorems (includes joining midpoints of two sides of a triangle, three sides of a triangle, the sides of a quadrilateral, median of a trapezoid; centroid, orthocenter, incenter, and circumcenter)	3, 18, 28	2 + 2 + 2 = 6		
16. Circles and Angle Measurement (includes ≅ tangents from same exterior point, radius ⊥ tangent, tangent circles, common tangents, arcs and chords, diameter ⊥ chord, arc length, area of a sector, center-radius equation, applying transformations)	9, 12, 25, 34	2 + 2 + 2 + 2 = 8		
17. Circles and Similar Triangles (includes proofs, segments of intersecting chords, tangent and secant segments)	17	2		
18. Area of Plane Figures (includes area of a regular polygon)	—	—		
19. Measurement of Solids (volume and lateral area; surface area of a sphere; great circle; comparing similar solids)	20, 30	2 + 2 = 4		

Map to Core Curriculum

Content Band	Item Numbers
Geometric Relationships	2, 8, 20, 30
Constructions	5, 32
Locus	28, 33
Informal and Formal Proofs	1, 3, 6, 7, 11, 12, 16, 17, 18, 22, 25, 26, 27, 29, 31, 35, 37
Transformational Geometry	15, 21, 23, 36
Coordinate Geometry	4, 9, 10, 13, 14, 19, 24, 34, 38

How to Convert Your Raw Score to Your Geometry Regents Examination Score

The conversion chart on the following page must be used to determine your final score on the August 2010 Regents Examination in Geometry. To find your final exam score, locate in the column labeled "Raw Score" the total number of points you scored out of a possible 86 points. Since partial credit is allowed in Parts II, III, and IV of the test, you may need to approximate the credit you would receive for a solution that is not completely correct. Then locate in the adjacent column to the right the scale score that corresponds to your raw score. The scale score is your final Geometry Regents Examination score.

Regents Examination in Geometry—August 2010

Chart for Converting Total Test Raw Scores to Final Examination Scores (Scale Scores)

Raw Score	Scale Score	Raw Score	Scale Score	Raw Score	Scale Score	Raw Score	Scale Score
86	100	64	80	42	67	20	43
85	98	63	79	41	66	19	41
84	97	62	79	40	65	18	40
83	96	61	78	39	64	17	38
82	94	60	78	38	63	16	36
81	93	59	77	37	62	15	35
80	92	58	77	36	62	14	33
79	91	57	76	35	61	13	31
78	90	56	75	34	60	12	29
77	89	55	75	33	59	11	26
76	88	54	74	32	58	10	24
75	87	53	74	31	57	9	22
74	87	52	73	30	56	8	20
73	86	51	73	29	55	7	17
72	85	50	72	28	54	6	15
71	84	49	71	27	52	5	12
70	84	48	71	26	51	4	10
69	83	47	70	25	50	3	7
68	82	46	69	24	49	2	5
67	82	45	69	23	47	1	2
66	81	44	68	22	46	0	0
65	80	43	67	21	45		

Examination
June 2011
Geometry

GEOMETRY REFERENCE SHEET

Volume	Cylinder	$V = Bh$, where B is the area of the base
	Pyramid	$V = \frac{1}{3}Bh$, where B is the area of the base
	Right circular cone	$V = \frac{1}{3}Bh$, where B is the area of the base
	Sphere	$V = \frac{4}{3}\pi r^3$

Lateral area (L)	Right circular cylinder	$L = 2\pi rh$
	Right circular cone	$L = \pi r\ell$, where ℓ is the slant height

Surface area	Sphere	$SA = 4\pi r^2$

PART I

Answer all 28 questions in this part. Each correct answer will receive 2 credits. No partial credit will be allowed. For each question, write in the space provided the number preceding the word or expression that best completes the statement or answers the question.   [56 credits]

1 Line segment AB is shown in the diagram below.

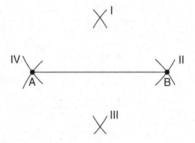

Which two sets of construction marks, labeled I, II, III, and IV, are part of the construction of the perpendicular bisector of line segment AB?

(1) I and II   (3) II and III
(2) I and III   (4) II and IV   1 _2_

2 If $\triangle JKL \cong \triangle MNO$, which statement is always true?

(1) $\angle KLJ \cong \angle NMO$ (3) $\overline{JL} \cong \overline{MO}$
(2) $\angle KJL \cong \angle MON$ (4) $\overline{JK} \cong \overline{ON}$ 2 _3_

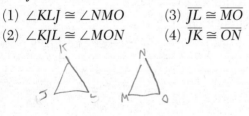

3 In the diagram below, $\triangle A'B'C'$ is a transformation of $\triangle ABC$, and $\triangle A''B''C''$ is a transformation of $\triangle A'B'C'$.

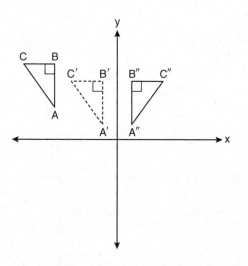

The composite transformation of $\triangle ABC$ to $\triangle A''B''C''$ is an example of a

(1) reflection followed by a rotation
(2) reflection followed by a translation
(3) translation followed by a rotation
(4) translation followed by a reflection

3 _4_

4 In the diagram below of $\triangle ACE$, medians $\overline{AD}$, $\overline{EB}$, and $\overline{CF}$ intersect at G. The length of $\overline{FG}$ is 12 cm.

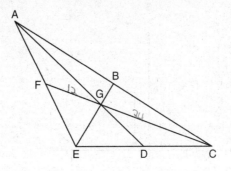

What is the length, in centimeters, of $\overline{GC}$?

(1) 24   (3) 6

(2) 12   (4) 4   4 __1__

5 In the diagram below of circle O, chord $\overline{AB}$ is parallel to chord $\overline{CD}$.

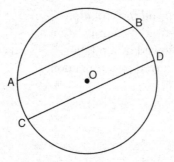

Which statement must be true?

(1) $\overset{\frown}{AC} \cong \overset{\frown}{BD}$ (3) $\overline{AB} \cong \overline{CD}$

(2) $\overset{\frown}{AB} \cong \overset{\frown}{CD}$ (4) $\overset{\frown}{ABD} \cong \overset{\frown}{CDB}$ 5 __1__

6 In the diagram below, line p intersects line m and line n.

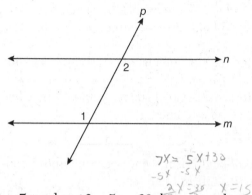

If m∠1 = 7x and m∠2 = 5x + 30, lines m and n are parallel when x equals

(1) 12.5   (3) 87.5
(2) 15   (4) 105   6 __2__

7 In the diagram of △KLM below, m∠L = 70, m∠M = 50, and $\overline{MK}$ is extended through N.

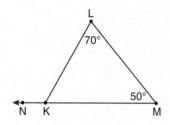

What is the measure of ∠LKN?

(1) 60°   (3) 180°
(2) 120°   (4) 300°   7 __2__

8 If two distinct planes, $\mathcal{A}$ and $\mathcal{B}$, are perpendicular to line c, then which statement is true?

(1) Planes $\mathcal{A}$ and $\mathcal{B}$ are parallel to each other.

(2) Planes $\mathcal{A}$ and $\mathcal{B}$ are perpendicular to each other.

(3) The intersection of planes $\mathcal{A}$ and $\mathcal{B}$ is a line parallel to line c.

(4) The intersection of planes $\mathcal{A}$ and $\mathcal{B}$ is a line perpendicular to line c.

8 __1__

9 What is the length of the line segment whose endpoints are $A(-1,9)$ and $B(7,4)$?

(1) $\sqrt{61}$ (3) $\sqrt{205}$

(2) $\sqrt{89}$ (4) $\sqrt{233}$

$\sqrt{(4-9)^2 + (7+1)^2}$

$25 + 64$

$\sqrt{89}$

9 __2__

10 What is an equation of circle O shown in the graph below?

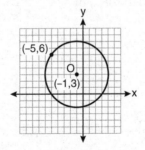

(1) $(x + 1)^2 + (y - 3)^2 = 25$
(2) $(x - 1)^2 + (y + 3)^2 = 25$
(3) $(x - 5)^2 + (y + 6)^2 = 25$
(4) $(x + 5)^2 + (y - 6)^2 = 25$

10 __1__

11 In the diagram below, parallelogram *ABCD* has diagonals $\overline{AC}$ and $\overline{BD}$ that intersect at point *E*.

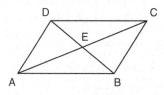

Which expression is *not* always true?

(1) $\angle DAE \cong \angle BCE$ (3) $\overline{AC} \cong \overline{DB}$

(2) $\angle DEC \cong \angle BEA$ (4) $\overline{DE} \cong \overline{EB}$ ✗11 __2__

12 The volume, in cubic centimeters, of a sphere whose diameter is 6 centimeters is

(1) 12π (3) 48π

(2) 36π (4) 288π 12 __2__

13 The equation of line *k* is $y = \frac{1}{3}x - 2$. The equation of line *m* is $-2x + 6y = 18$. Lines *k* and *m* are

(1) parallel
(2) perpendicular
(3) the same line
(4) neither parallel nor perpendicular   13 __1__

$$6y = 2x + 18$$
$$\frac{6y}{6} = \frac{2x}{6} + \frac{18}{6}$$

$$y = \frac{1}{3}x + 3$$

14 What are the center and the radius of the circle whose equation is $(x - 5)^2 + (y + 3)^2 = 16$?

(1) $(-5,3)$ and 16   (3) $(-5,3)$ and 4

(2) $(5,-3)$ and 16   (4) $(5,-3)$ and 4   14 __4__

$+5, -3 = 4$

15 Triangle ABC has vertices $A(0,0)$, $B(3,2)$, and $C(0,4)$. This triangle may be classified as

(1) equilateral   (3) right

(2) isosceles   (4) scalene   15 __2__

$\sqrt{(4-2)^2 (0-3)^2}$ $\sqrt{(4-0)^2 (2-0)^2}$ $\sqrt{(2-0)^2 (3-0)^2}$

$4 + \sqrt{13}$ $\sqrt{16}$ $4 + 9 \sqrt{13}$

16 In rhombus $ABCD$, the diagonals $\overline{AC}$ and $\overline{BD}$ intersect at E. If $AE = 5$ and $BE = 12$, what is the length of $\overline{AB}$?

(1) 7   (3) 13

(2) 10   (4) 17   ✗16 __4__

17 In the diagram below of circle O, $\overline{PA}$ is tangent to circle O at A, and $\overline{PBC}$ is a secant with points B and C on the circle.

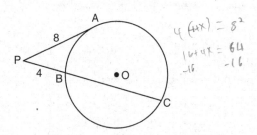

$4(4+x) = 8^2$

$16 + 4x = 64$

$-16 \qquad -16$

If $PA = 8$ and $PB = 4$, what is the length of $\overline{BC}$?

(1) 20   (3) 15

(2) 16   (4) 12   17 __4__

18 Lines *m* and *n* intersect at point *A*. Line *k* is perpendicular to both lines *m* and *n* at point *A*. Which statement *must* be true?

(1) Lines *m*, *n*, and *k* are in the same plane.
(2) Lines *m* and *n* are in two different planes.
(3) Lines *m* and *n* are perpendicular to each other.
(4) Line *k* is perpendicular to the plane containing lines *m* and *n*.

18 ___4___

19 In △*DEF*, m∠*D* = 3*x* + 5, m∠*E* = 4*x* − 15, and m∠*F* = 2*x* + 10. Which statement is true?

(handwritten: 65 above m∠D, 65 above m∠E, 50 above m∠F)

(1) *DF* = *FE*   (3) m∠*E* = m∠*F*
(2) *DE* = *FE*   (4) m∠*D* = m∠*F*

19 ___1___

(handwritten work: 9x = 180, x = 20; 3x+5 +4x−15 +2x+10 = 180)

20 As shown in the diagram below, △*ABC* ~ △*DEF*, *AB* = 7*x*, *BC* = 4, *DE* = 7, and *EF* = *x*.

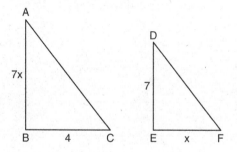

What is the length of $\overline{AB}$?

(1) 28   (3) 14
(2) 2   (4) 4

20 ___3___

21 A man wants to place a new bird bath in his yard so that it is 30 feet from a fence, *f*, and also 10 feet from a light pole, *P*. As shown in the diagram below, the light pole is 35 feet away from the fence.

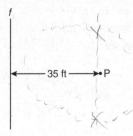

How many locations are possible for the bird bath?

(1) 1   (3) 3

(2) 2   (4) 0   21 __2__

22 As shown on the graph below, $\triangle R'S'T'$ is the image of $\triangle RST$ under a single transformation.

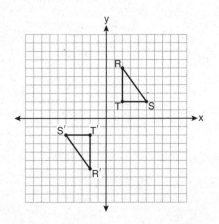

Which transformation does this graph represent?

(1) glide reflection   (3) rotation

(2) line reflection   (4) translation   22 __2__

23 Which line is parallel to the line whose equation is $4x + 3y = 7$ and also passes through the point $(-5,2)$?

(1) $4x + 3y = -26$ (3) $3x + 4y = -7$

(2) $4x + 3y = -14$ (4) $3x + 4y = 14$ 23 ___2___

24 If the vertex angles of two isosceles triangles are congruent, then the triangles must be

(1) acute   (3) right

(2) congruent   (4) similar   24 ___2___

25 Which quadrilateral has diagonals that always bisect its angles and also bisect each other?

(1) rhombus   (3) parallelogram

(2) rectangle   (4) isosceles trapezoid   25 ___2___

26 When $\triangle ABC$ is dilated by a scale factor of 2, its image is $\triangle A'B'C'$. Which statement is true?

(1) $\overline{AC} \cong \overline{A'C'}$

(2) $\angle A \cong \angle A'$

(3) perimeter of $\triangle ABC$ = perimeter of $\triangle A'B'C'$

(4) 2(area of $\triangle ABC$) = area of $\triangle A'B'C'$ 26 ___4___

27 What is the slope of a line that is perpendicular to the line whose equation is $3x + 5y = 4$?

(1) $-\dfrac{3}{5}$ (3) $-\dfrac{5}{3}$

(2) $\dfrac{3}{5}$ (4) $\dfrac{5}{3}$

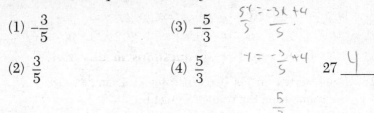

$$\frac{5y = -3x + 4}{5}$$

$$y = \frac{-3}{5} + 4$$

27 ___4___

$$\frac{5}{3}$$

28 In the diagram below of right triangle ABC, altitude $\overline{BD}$ is drawn to hypotenuse $\overline{AC}$, $AC = 16$, and $CD = 7$.

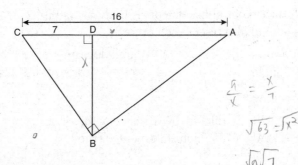

$$\frac{9}{x} = \frac{x}{7}$$

$$\sqrt{63} = \sqrt{x^2}$$

$$\sqrt{9}\sqrt{7}$$

$$3\sqrt{7}$$

What is the length of $\overline{BD}$?

(1) $3\sqrt{7}$ (3) $7\sqrt{3}$

(2) $4\sqrt{7}$ (4) 12

28 ___1___

PART II

Answer all 6 questions in this part. Each correct answer will receive 2 credits. Clearly indicate the necessary steps, including appropriate formula substitutions, diagrams, graphs, charts, etc. For all questions in this part, a correct numerical answer with no work shown will receive only 1 credit.   [12 credits]

29 Given the true statement, "The medians of a triangle are concurrent," write the negation of the statement and give the truth value for the negation.

The medians of a triangle are not concurrent

false

30 Using a compass and straightedge, on the diagram

below of $\overrightarrow{RS}$, construct an equilateral triangle with

$\overline{RS}$ as one side. [Leave all construction marks.]

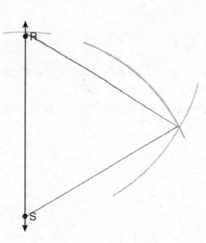

31 The Parkside Packing Company needs a rectangular shipping box. The box must have a length of 11 inches and a width of 8 inches. Find, to the *nearest tenth of an inch*, the minimum height of the box such that the volume is *at least* 800 cubic inches.

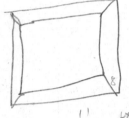

11   6×1

L×w×H

88   8×11 = 8

88 × 10 = 880

11×8×h ≥ 800

$$\frac{88h}{88} \geq \frac{800}{88}$$

h ≥

h ≥ 9.09

9.1

32 A pentagon is drawn on the set of axes below. If the pentagon is reflected over the y-axis, determine if this transformation is an isometry.

Justify your answer. [The use of the set of axes below is optional.]

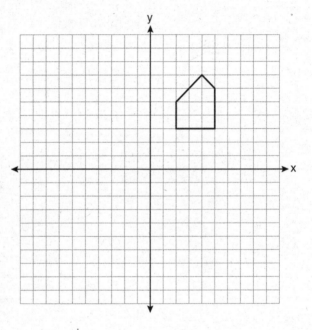

Yes since Distance and slopes of the two angles are congruent

33 In the diagram below of $\triangle ABC$, D is a point on $\overline{AB}$, E is a point on $\overline{BC}$, $\overline{AC} \parallel \overline{DE}$, $CE = 25$ inches, $AD = 18$ inches, and $DB = 12$ inches. Find, to the *nearest tenth of an inch*, the length of $\overline{EB}$.

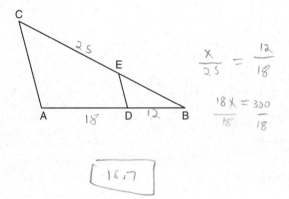

$$\frac{x}{25} = \frac{12}{18}$$

$$\frac{18x}{18} = \frac{300}{18}$$

$$\boxed{16.7}$$

34 In circle O, diameter $\overline{RS}$ has endpoints $R(3a, 2b - 1)$ and $S(a - 6, 4b + 5)$. Find the coordinates of point O, in terms of a and b. Express your answer in simplest form.

$$\frac{3a + a - 6}{2} \qquad \frac{2b - 1 + 4b + 5}{2}$$

$$\qquad\qquad\qquad \frac{6b + 4}{2}$$

$$\frac{4a - 6}{2} \qquad \boxed{2a - 3, \; 3b + 2}$$

PART III

Answer all 3 questions in this part. Each correct answer will receive 4 credits. Clearly indicate the necessary steps, including appropriate formula substitutions, diagrams, graphs, charts, etc. For all questions in this part, a correct numerical answer with no work shown will receive only 1 credit. [12 credits]

35 On the set of coordinate axes below, graph the locus of points that are equidistant from the lines $y = 6$ and $y = 2$ and also graph the locus of points that are 3 units from the y-axis. State the coordinates of *all* points that satisfy *both* conditions.

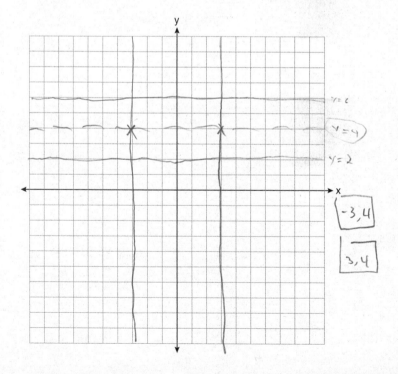

36 In the diagram below, tangent $\overline{ML}$ and secant $\overline{MNK}$ are drawn to circle O. The ratio $m\stackrel{\frown}{LN}{:}m\stackrel{\frown}{NK}{:}m\stackrel{\frown}{KL}$ is 3:4:5. Find $m\angle LMK$.

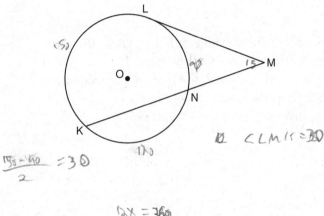

$\frac{150-90}{2}=30$

$\dfrac{12x}{12}=\dfrac{360}{12}$

$x=30$

7 Solve the following system of equations graphically.

$$2x^2 - 4x = y + 1$$
$$x + y = 1$$

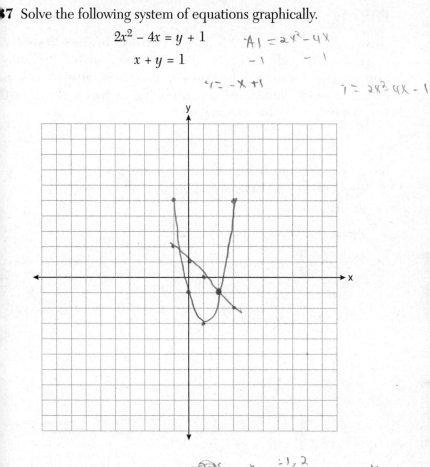

$y = 2x^2 - 4x$
$-1 \qquad -1$
$y = -x + 1$
$y = 2x^2 - 4x - 1$

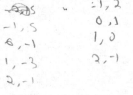

$-1, 5$ $-1, 2$
$0, -1$ $0, 1$
$1, -3$ $1, 0$
$2, -1$ $2, -1$

$2, -1$ ans

PART IV

Answer the question in this part. A correct answer will receive 6 credits. Clearly indicate the necessary steps, including appropriate formula substitutions, diagrams, graphs, charts, etc. A correct numerical answer with no work shown will receive only 1 credit.   [6 credits]

38 In the diagram below, $\overline{PA}$ and $\overline{PB}$ are tangent to circle O, $\overline{OA}$ and $\overline{OB}$ are radii, and $\overline{OP}$ intersects the circle at C.

Prove: $\angle AOP \cong \angle BOP$

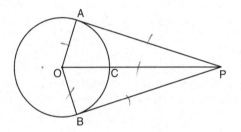

$AP \cong PB$	tangent from same spot are consequent
$AO \cong OB$	radii are equal
$OP \cong OP$	reflexive prop
$\triangle OAP \cong \triangle OBP$	$SSS \cong SSS$
$\angle AOP \cong \angle BOP$	CPCTC

Answers
June 2011
Geometry

Answer Key

PART I

1. (2)	**8.** (1)	**15.** (2)	**22.** (3)
2. (3)	**9.** (2)	**16.** (3)	**23.** (2)
3. (4)	**10.** (1)	**17.** (4)	**24.** (4)
4. (1)	**11.** (3)	**18.** (4)	**25.** (1)
5. (1)	**12.** (2)	**19.** (1)	**26.** (2)
6. (2)	**13.** (1)	**20.** (3)	**27.** (4)
7. (2)	**14.** (4)	**21.** (2)	**28.** (1)

PART II

29. See the *Answers Explained* section.

30. See the *Answers Explained* section.

31. 9.1

32. Yes, the line reflection is an isometry.

33. 16.7

34. $(2a - 3, 3b + 2)$

PART III

35. $(-3,4)$ and $(3,4)$

36. 30

37. $(2,-1)$ and $(-0.5,1.5)$

PART IV

38. See the *Answers Explained* section.

In **PARTS II–IV** you are required to show how you arrived at your answers. For sample methods of solutions, see the *Answers Explained* section.

Answers Explained

PART I

1. You are asked to determine the set of construction marks in the accompanying diagram that are part of the construction of the perpendicular bisector of line segment *AB*.

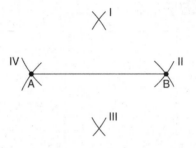

To construct the perpendicular bisector of $\overline{AB}$, arcs with the same radius must be drawn above and below the segment using each endpoint as a center, as illustrated by the pair of arcs labeled I and III.

The correct choice is **(2)**.

2. It is given that $\triangle JKL \cong \triangle MNO$. You are asked to identify from among the answer choices a pair of congruent parts. Draw two congruent triangles and label their vertices such that $J \leftrightarrow M$, $K \leftrightarrow N$, and $L \leftrightarrow O$, as shown in the accompanying diagram.

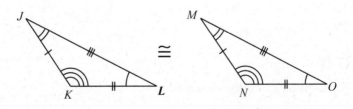

Mark off corresponding parts as congruent. Then examine each answer choice in turn.

- Choice (1): $\angle KLJ \cong \angle NMO$. This is not always true since angles L and M are not corresponding angles of congruent triangles. ✗

- Choice (2): $\angle KJL \cong \angle MON$. This is not always true since angles J and O are not corresponding angles of congruent triangles. ✗

- Choice (3): $\overline{JL} \cong \overline{MO}$. This is always true since these sides are corresponding sides of congruent triangles. ✔

- Choice (4): $\overline{JK} \cong \overline{ON}$. This is not always true since these sides are not corresponding sides of congruent triangles. ✗

The correct choice is **(3)**.

3. In the accompanying diagram, $\triangle A'B'C'$ is a transformation of $\triangle ABC$, and $\triangle A''B''C''$ is a transformation of $\triangle A'B'C'$.

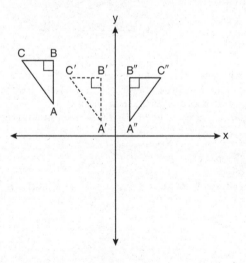

You are asked to identify from among the answer choices the type of composite transformation that maps $\triangle ABC$ to $\triangle A''B''C''$.

The transformation that maps $\triangle ABC$ to $\triangle A'B'C'$ is a translation since $\triangle ABC$ is simply shifted down and to the right.

The transformation that maps △A'B'C' to △A"B"C" is a reflection since △A"B"C" is produced by flipping or reflecting △A'B'C' over the y-axis.

Hence, the composite transformation of △ABC to △A"B"C" is an example of a translation followed by a reflection.

The correct choice is **(4)**.

4. In the accompanying diagram of △ACE, medians $\overline{AD}$, $\overline{EB}$, and $\overline{CF}$ intersect at G. Given that the length of $\overline{FG}$ is 12 cm, you must find the length of $\overline{GC}$.

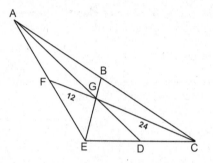

The point at which the three medians of a triangle intersect divides each median into two segments whose lengths are in the ratio 2:1, with the longest segment having a vertex of the triangle as an endpoint. Since GC:FG = 2:1, GC must be two times as great as FG. If FG =12 cm, then C = 2 × 12 cm = 24 cm.

The correct choice is **(1)**.

5. In the accompanying diagram of circle O, chord $\overline{AB}$ is parallel to chord $\overline{CD}$.

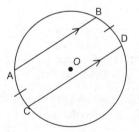

Since parallel chords in a circle intercept congruent arcs, $\overset{\frown}{AC} \cong \overset{\frown}{BD}$.

The correct choice is (**1**).

6. It is given that in the accompanying diagram, line p intersects line m and line n.

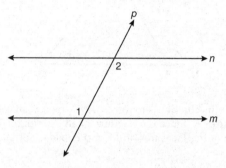

It is also given that m$\angle 1$ = $7x$ and m$\angle 2$ = $5x + 30$. You are asked to find the value of x for which lines m and n are parallel. Angles 1 and 2 are alternate interior angles. Lines m and n are parallel when alternate interior angles are equal in measure. To find the value of x that makes lines m and n parallel, set m$\angle 1$ equal to m$\angle 2$ and solve for x:

$$m\angle 1 = m\angle 2$$
$$7x = 5x + 30$$
$$7x - 5x = 5x + 30 - 5x$$
$$2x = 30$$
$$\frac{2x}{2} = \frac{30}{2}$$
$$x = 15$$

The correct choice is (**2**).

7. In the accompanying diagram of $\triangle KLM$, $m\angle L = 70$, $m\angle M = 50$, and $\overline{MK}$ is extended through N.

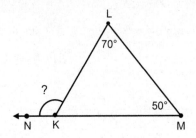

You are asked to find the measure of $\angle LKN$, which is an exterior angle of the triangle with vertex K. Since the measure of an exterior angle of a triangle is equal to the sum of the measures of the two nonadjacent interior angles of the triangle:

$$m\angle LKN = m\angle L + m\angle M$$
$$= 70 + 50$$
$$= 120$$

The correct choice is (**2**).

8. Given that two distinct planes, $\mathcal{A}$ and $\mathcal{B}$, are perpendicular to line c, you are asked to determine which of the answer choices contains a true statement. If two different planes are each perpendicular to the same line, the two planes must be parallel to each other, as illustrated in the accompanying diagram.

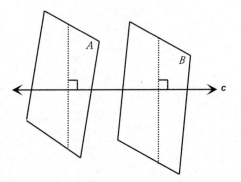

The correct choice is (**1**).

9. To find the length, d, of the line segment whose endpoints are $A(-1,9)$ nd $B(7,4)$, use the distance formula:

$$d = \sqrt{(\Delta x)^2 + (\Delta y)^2}$$

In this formula, Δx represents the difference in the x-coordinates of the wo given points and Δy represents the difference in the y-coordinates, with he subtractions performed in the same order:

$$
\begin{aligned}
AB &= \sqrt{(7-(-1))^2 + (4-9)^2} \\
&= \sqrt{(8)^2 + (-5)^2} \\
&= \sqrt{64 + 25} \\
&= \sqrt{89}
\end{aligned}
$$

The correct choice is **(2)**.

10. You are asked to determine an equation of the circle shown in the ccompanying graph.

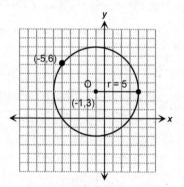

The center of circle O is at $(-1,3)$. To determine the radius of the ircle, count the number of unit boxes from point O horizontally across to he circle. The radius of the circle is 5. The center-radius form of the equation f a circle is $(x - h)^2 + (y - k)^2 = r^2$, where (h,k) is the center of the ircle and r is the radius. Substituting -1 for h, 3 for k, and 5 for r gives $x - (-1))^2 + (y - 3)^2 = 5^2$ or $(x + 1)^2 + (y - 3)^2 = 25$.

The correct choice is **(1)**.

11. It is given that in parallelogram $ABCD$, diagonals $\overline{AC}$ and $\overline{BD}$ intersect at point E. To determine which answer choice is *not* always true, examine each choice in turn.

- Choice (1): $\angle DAE \cong \angle BCE$. Since opposite sides of a parallelogram are parallel, $\overline{AD} \parallel \overline{BC}$. Since $\angle DAE$ and $\angle BCE$ are alternate interior angles, they are congruent. Hence, the expression $\angle DAE \cong \angle BCE$ is always true.

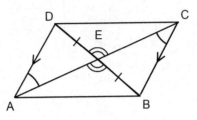

- Choice (2): $\angle DEC \cong \angle BEA$. Since vertical angles are congruent, this expression is always true.

- Choice (3): $\overline{AC} \cong \overline{DB}$. Since the diagonals of a parallelogram are not necessarily congruent, this expression is *not* always true.

- Choice (4): $\overline{DE} \cong \overline{EB}$. The diagonals of a parallelogram bisect each other, so this expression is always true.

The correct choice is **(3)**.

12. To find the volume, in cubic centimeters, of a sphere whose diameter is 6 centimeters, use the formula $V = \frac{4}{3}\pi r^3$, where r is the radius of a sphere whose volume is V. This formula is shown on the reference sheet. Since the diameter of the sphere is 6 cm, its radius is 3 cm:

$$V = \frac{4}{3}\pi(3)^3$$

$$= \frac{4}{\cancel{3}}\pi(\cancel{27}^{\,9})$$

$$= 36\pi$$

The correct choice is **(2)**.

13. The equation of line k is given as $y = \frac{1}{3}x - 2$, and the equation of line m is given as $-2x + 6y = 18$. To determine the relationship, if any, between the two lines, compare the equations of the lines when each is written in the slope-intercept $y = mx + b$ form where the coefficient of x represents the slope of the line and b is its y-intercept.

Write the equation of line m in slope-intercept form by solving for y:

$$-2x + 6y = 18$$

$$6y = 2x + 18$$

$$\frac{6y}{6} = \frac{2x}{6} + \frac{18}{6}$$

$$y = \frac{1}{3}x + 3$$

Since the equation of line k is given as $y = \frac{1}{3}x - 2$, the two lines have different y-intercepts. So they cannot be the same line.

Lines k and m have the same slope, $\frac{1}{3}$, and are therefore parallel.

The correct choice is **(1)**.

14. To determine the center and radius of the circle whose equation is given as $(x - 5)^2 + (y + 3)^2 = 16$, rewrite the equation in the center-radius form $(x - h)^2 + (y - k)^2 = r^2$, where (h,k) is the center of the circle and r is the radius:

$$\left(x - \underset{h}{\underbrace{5}}\right)^2 + \left(y - \underset{k}{\underbrace{(-3)}}\right)^2 = \underset{r}{\underbrace{4}}^2$$

Since $(h,k) = (5,-3)$ and $r = 4$, the center of the circle is at $(5,-3)$ and its radius is 4.

The correct choice is **(4)**.

15. The coordinates of the vertices of triangle *ABC* are given as *A*(0,0), *B*(3,2), and *C*(0,4). Sketch the triangle:

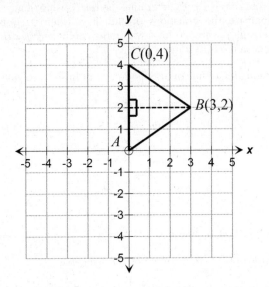

Since drawing an altitude from *B* to $\overline{AC}$ forms two congruent right triangles with $\overline{AB} \cong \overline{BC}$, triangle *ABC* may be classified as isosceles.

The correct choice is (**2**).

16. Diagonals $\overline{AC}$ and $\overline{BD}$ of rhombus *ABCD* intersect at *E* with *AE* = 5 and *BE* = 12. To find the length of $\overline{AB}$, use the fact that the diagonals of a rhombus intersect at right angles:

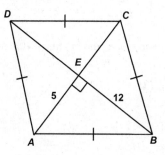

If x represents the length of $\overline{AB}$, then according to the Pythagorean theorem:

$$x^2 = (5)^2 + (12)^2$$
$$= 25 + 144$$
$$= 169$$
$$\sqrt{x^2} = \sqrt{169}$$
$$x = 13$$

The length of $\overline{AB}$ is 13.

The correct choice is **(3)**.

17. In the accompanying diagram of circle O, $\overline{PA}$ is tangent to circle O at A, $PA = 8$, $\overline{PBC}$ is a secant with points B and C on the circle, and $PB = 4$. You are asked to find the length of $\overline{BC}$.

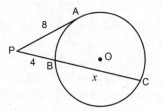

If x represents the length of $\overline{BC}$, then the length of secant $\overline{PBC}$ is $x + 4$. If a tangent and a secant are drawn to a circle from the same exterior point, the length of the tangent segment is the mean proportional between the lengths of the external segment of the secant and the whole secant:

$$\frac{PB}{PA} = \frac{PA}{PBC}$$
$$\frac{4}{8} = \frac{8}{x+4}$$
$$4(x + 4) = 8 \cdot 8$$
$$4x + 16 = 64$$
$$4x + 16 - 16 = 64 - 16$$
$$4x = 48$$
$$\frac{4x}{4} = \frac{48}{4}$$
$$x = 12$$

The length of $\overline{BC}$ is 12.

The correct choice is **(4)**.

18. You are asked to determine the statement that is true given that lines m and n intersect at point A and that line k is perpendicular to both lines at point A. If a line is perpendicular to each of two intersecting lines at their point of intersection, the line is perpendicular to the plane determined by them, as illustrated in the accompanying diagram.

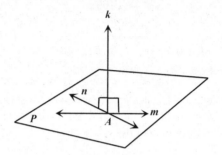

If $k \perp m$ and $k \perp n$, then $k \perp$ plane P.

The correct choice is **(4)**.

19. In $\triangle DEF$, $m\angle A = 3x + 5$, $m\angle E = 4x - 15$, and $m\angle F = 2x + 10$. Since the sum of the measures of the angles of a triangle is 180:

$$(3x + 5) + (4x - 15) + (2x + 10) = 180$$

$$(3x + 4x + 2x) + (5 - 15 + 10) = 180$$

$$9x = 180$$

$$\frac{9x}{9} = \frac{180}{9}$$

$$x = 20$$

Substitute $x = 20$ into each expression to find the measure of each angle.

$$m\angle D = 3x + 5$$
$$= 3(20) + 5$$
$$= 65$$

$$m\angle E = 4x - 15$$
$$= 4(20) - 15$$
$$= 65$$

$$m\angle F = 2x + 10$$
$$= 2(20) + 10$$
$$= 50$$

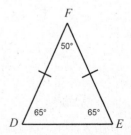

If two angles of a triangle are equal in measure, then the sides opposite those angles are equal in length. Hence, $DF = FE$.

The correct choice is **(1)**.

20. As shown in the accompanying diagram, $\triangle ABC \sim \triangle DEF$, $AB = 7x$, $BC = 4$, $DE = 7$, and $EF = x$.

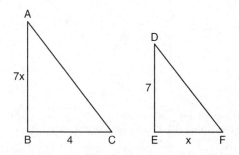

You are asked to find the length of $\overline{AB}$. Since the lengths of corresponding sides of similar triangles are in proportion:

$$\frac{AB}{DE} = \frac{BC}{EF}$$

$$\frac{7x}{7} = \frac{4}{x}$$

$$\frac{\cancel{7}x}{{}_1\cancel{7}} = \frac{4}{x}$$

$$x \cdot x = 4 \cdot 1$$

$$x^2 = 4$$

$$x = \sqrt{4} = 2$$

Hence, $AB = 7x = 7(2) = 14$.

The correct choice is **(3)**.

21. A man wants to place a new bird bath in his yard so that it is 30 feet from a fence, f, and also 10 feet from a light pole P. As shown in the accompanying diagram, the light pole is 35 feet away from the fence.

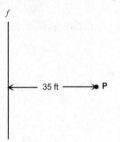

To find the number of possible locations for the bird bath, determine the number of points at which the two locus conditions intersect.

The locus of points 30 feet from fence f are parallel lines with each of the parallel lines 30 feet from f:

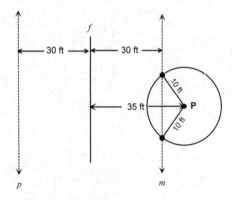

The locus of points 10 feet from light pole P is a circle with P as its center and a radius of 10 feet.

Because the broken line labeled m is 30 feet from f, it is $35 - 30 = 5$ feet from P. Since the radius of the circle is 10 feet, the circle intersects line m in 2 different points, as shown in the accompanying diagram.

The correct choice is **(2)**.

22. It is given that on the accompanying graph, $\triangle R'S'T'$ is the image of $\triangle RST$ under a single transformation.

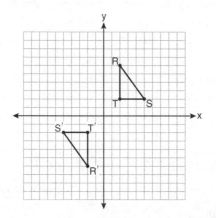

To identify the transformation that this graph represents, consider each answer choice in turn.

- Choice (1): glide reflection. A glide reflection is the composition of a reflection in a line and a translation in a direction along that line. Reflecting $\triangle RST$ in either coordinate axis and then shifting the image in a direction parallel to that axis does not produce $\triangle R'S'T'$. ✗

- Choice (2): line reflection. Reflecting $\triangle RST$ in either coordinate axis does not produce $\triangle R'S'T'$. ✗

- Choice (3): rotation. Since the origin is the midpoint of $\overline{RR'}$, $\overline{SS'}$, and $\overline{TT'}$, points R', S', and T' are the images of points R, S, and T, respectively, under a point reflection in the origin. Because a point reflection in the origin and a rotation of 180° about the origin are equivalent transformations, the graph represents a rotation of $\triangle RST$ 180° about the origin. ✔

- Choice (4): translation. Shifting $\triangle RST$ horizontally, vertically, or both horizontally and vertically does not produce $\triangle R'S'T'$. ✗

The correct choice is **(3)**.

23. You are asked to identify the equation of the line that is parallel to the line whose equation is $4x + 3y = 7$ and also passes through the point $(-5,2)$. First determine the slope of the given line by writing its equation in slope-intercept form, $y = mx + b$, where m is the slope of the line and b is its y-intercept:

$$4x + 3y = 7$$

$$3y = -4x + 7$$

$$\frac{3y}{3} = \frac{-4x}{3} + \frac{7}{3}$$

$$y = -\frac{4}{3}x + \frac{7}{3}$$

Hence, the slope of the given line is $-\frac{4}{3}$, the coefficient of x.

Since parallel lines have the same slope, the slope of the required line is also $-\frac{4}{3}$.

The equation of the required line has the form $y = -\frac{4}{3}x + b$. Since the point $(-5,2)$ is on the line, its coordinates must satisfy the equation of the line:

$$2 = -\frac{4}{3}(-5) + b$$

$$2 = \frac{20}{3} + b$$

$$b = 2 - \frac{20}{3}$$

$$= \frac{6}{3} - \frac{20}{3}$$

$$= -\frac{14}{3}$$

An equation of the required line is $y = -\frac{4}{3}x - \frac{14}{3}$. Write this equation in the form $Ax + By = C$ so that it matches the form of the answer choices. Eliminate the fractional terms of the equation by multiplying each term of the equation by 3. Then collect variable terms on the left side of the equation:

$$3(y) = 3\left(-\frac{4}{\overset{3}{1}}x\right) - 3\left(\frac{14}{\overset{3}{1}}\right)$$

$$3y = -4x - 14$$

$$4x + 3y = -14$$

The correct choice is (**2**).

24. If the vertex angles of two isosceles triangles are congruent, then their base angles must also be congruent.

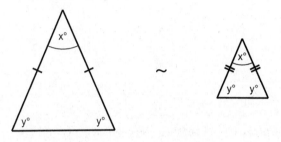

Since the three angles of one triangle are congruent to the corresponding angles of the second triangle, the two triangles must be similar.

The correct choice is (**4**).

25. A rhombus is a parallelogram with four congruent sides. The diagonals of a rhombus bisect each other (as in all parallelograms) and also bisect its angles.

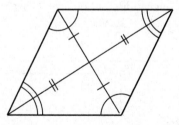

The correct choice is (**1**).

26. When $\triangle ABC$ is dilated by a scale factor of 2, its image is $\triangle A'B'C'$. Under this dilation, the length of each side of $\triangle ABC$ is multiplied by 2 while the measure of each angle remains the same. Hence, corresponding angles of the two triangles must be congruent. In particular, $\angle A \cong \angle A'$.

The correct choice is **(2)**.

27. To find the slope of a line that is perpendicular to the line whose equation is $3x + 5y = 4$, first write the given equation in slope-intercept form, $y = mx + b$, where m is the slope of the line and b is its y-intercept:

$$3x + 5y = 4$$

$$5y = -3x + 4$$

$$\frac{5y}{5} = \frac{-3x}{5} + \frac{4}{5}$$

$$y = -\frac{3}{5}x + \frac{4}{5}$$

The slope of the given line is $-\frac{3}{5}$, the coefficient of x.

Perpendicular lines have slopes that are negative reciprocals. The negative reciprocal of $-\frac{3}{5}$ is $\frac{5}{3}$ since $-\frac{3}{5} \times \frac{5}{3} = -1$.

Hence, a line with a slope of $\frac{5}{3}$ is perpendicular to the given line.

The correct choice is **(4)**.

28. In the accompanying diagram of right triangle ABC, altitude $\overline{BD}$ is drawn to hypotenuse $\overline{AC}$, $AC = 16$, and $CD = 7$.

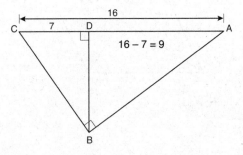

If an altitude is drawn to the hypotenuse of a right triangle, the length of the altitude is the mean proportional between the lengths of segments of the hypotenuse:

$$\frac{CD}{BD} = \frac{BD}{AD}$$

Since AC is formed by CD and AD, $AD = AC - CD = 16 - 7 = 9$. Thus:

$$\frac{7}{BD} = \frac{BD}{9}$$

$$(BD)^2 = 9 \cdot 7$$

$$BD = \sqrt{9 \cdot 7}$$

$$= \sqrt{9} \cdot \sqrt{7}$$

$$= 3\sqrt{7}$$

The correct choice is (**1**).

PART II

29. Given the true statement, "The medians of a triangle are concurrent," you must write the negation of the statement and give the truth value of the negation. To form the negation of a statement, write the statement that has the opposite truth value. The negation of the given statement is, "The medians of a triangle are not concurrent." The negation must be false since the original statement is true.

30. Using a compass, straightedge, and the diagram below, you are required to construct an equilateral triangle with $\overline{RS}$ as one side.

Plan: Set the radius length of your compass equal to the length of $\overline{RS}$. Use this compass setting to draw arcs using points R and S as centers. Label the point of intersection of the two arcs point T.

- Step 1: Set the radius length of your compass equal to the length of $\overline{RS}$ by placing the fixed point of the compass on R and the pencil point on S.

- Step 2: Using this compass setting and keeping the compass point on R, draw an arc.

- Step 3: Place the compass point on S while keeping the compass setting fixed. Draw an arc that intersects the first arc. Label the point of intersection as T.

- Step 4: Using a straightedge, draw $\overline{RT}$ and $\overline{ST}$. Since $\overline{RS} = \overline{RT} = \overline{ST}$, $\triangle RST$ is an equilateral triangle.

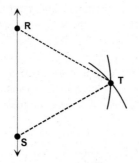

31. It is given that the Parkside Packing Company needs a rectangular shipping box. The box must have a length of 11 inches and a width of 8 inches. You are asked to find, to the *nearest tenth of an inch*, the minimum height of the box such that the volume is *at least* 800 cubic inches. The volume of the box is the product of its length, width, and height. Assume h represents the height of the box. Since the volume of the box must be greater than or equal to 800 cubic inches:

$$\underbrace{11 \times 8 \times h}_{\text{volume}} \geq 800$$

$$88h \geq 800$$

$$\frac{\cancel{88}h}{\cancel{88}} \geq \frac{880}{88}$$

$$h \geq 9.090909091$$

The minimum height of the box, to the *nearest tenth of an inch*, is **9.1** inches.

32. A pentagon is drawn on a set of coordinate axes and reflected over the y-axis. You are asked to determine if this transformation is an isometry. A transformation is an isometry if it preserves the distance between points so that the figure (preimage) and its image are congruent. Every reflection in a line preserves the distance between points and is therefore an isometry. Consequently, the transformation described must be an isometry.

33. In the accompanying diagram of $\triangle ABC$, D is a point on $\overline{AB}$, E is a point on $\overline{BC}$, $\overline{AC} \parallel \overline{DE}$, CE = 25 inches, AD = 18 inches, and DB = 12 inches. You are asked to find, to the *nearest tenth of an inch*, the length of $\overline{EB}$.

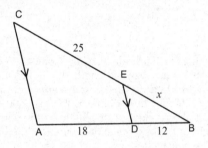

If a line segment whose endpoints are on two sides of a triangle is parallel to the third side of the triangle, that line segment divides the two sides proportionally. If x represents the length of $\overline{EB}$:

$$\frac{EB}{CE} = \frac{DB}{AD}$$

$$\frac{x}{25} = \frac{12}{18}$$

$$18x = 12 \times 25$$

$$18x = 300$$

$$\frac{\cancel{18}x}{\cancel{18}} = \frac{300}{18}$$

$$x \approx 16.66666667$$

The length of $\overline{EB}$, to the *nearest tenth of an inch*, is **16.7**.

34. In circle O, diameter $\overline{RS}$ has endpoints $R(3a, 2b-1)$ and $S(a-6, 4b+5)$. You are asked to find the coordinates of point O, in terms of a and b. Since the center of a circle is the midpoint of each diameter of the circle, use the midpoint formula to find the coordinates of point O. If the coordinates of point O are $(\bar{x}, \bar{y})$:

$$\bar{x} = \frac{3a + (a-6)}{2} \quad \text{and} \quad \bar{y} = \frac{(2b-1) + (4b+5)}{2}$$

$$= \frac{4a-6}{2} \qquad\qquad\qquad = \frac{6b+4}{2}$$

$$= \frac{4a}{2} - \frac{6}{2} \qquad\qquad\qquad = \frac{6b}{2} + \frac{4}{2}$$

$$= 2a - 3 \qquad\qquad\qquad\qquad = 3b + 2$$

The coordinates of point O are $(\textbf{2a} - \textbf{3, 3b} + \textbf{2})$.

PART III

35. On the set of coordinate axes, you are asked to graph the locus of points that are equidistant from the lines $y = 6$ and $y = 2$. You must also graph the locus of points that are 3 units from the y-axis and then state the coordinates of all points that satisfy both conditions.

The lines $y = 6$ and $y = 2$ are parallel lines. The locus of points equidistant between two parallel lines is a parallel line midway between them. Hence, the locus of points equidistant from the given lines is the line $y = \dfrac{6+2}{2} = \dfrac{8}{2} = 4$, as shown in the accompanying diagram.

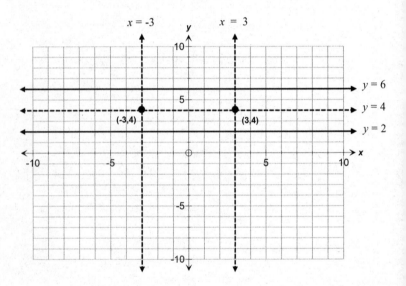

The locus of points 3 units from the y-axis is a pair of lines on opposite sides of the y-axis and each 3 units from it, as shown in the accompanying diagram. The equations of these lines are $x = -3$ and $x = 3$.

The points where both graphs intersect represent the points that satisfy both locus conditions.

Two points satisfy both locus conditions: **(–3,4)** and **(3,4)**.

36. In the accompanying diagram, tangent $\overline{ML}$ and secant $\overline{MNK}$ are drawn to circle O. The ratio m $\overset{\frown}{LN}$:m $\overset{\frown}{NK}$:m $\overset{\frown}{KL}$ is 3:4:5. You are asked to find m$\angle LMK$.

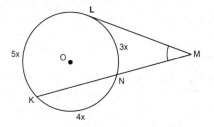

Since the sum of the measures of the arcs that comprise a circle is 360:

$$3x + 4x + 5x = 360$$
$$12x = 360$$
$$\frac{12x}{12x} = \frac{360}{12x}$$
$$x = 30$$

Use this information to find the measure of each arc.

$$\text{m } \overset{\frown}{LN} = 3x = 3(30) = 90$$

$$\text{m } \overset{\frown}{NK} = 4x = 4(30) = 120$$

$$\text{m } \overset{\frown}{KL} = 5x = 5(30) = 150$$

Angle LMK is formed by a tangent and a secant drawn to a circle from the same exterior point. As a result, angle LMK is equal in measure to one-half the difference of the measures of the intercepted arcs:

$$\text{m}\angle LMK = \frac{1}{2}(\text{m } \overset{\frown}{KL} - \text{m } \overset{\frown}{LN})$$

$$= \frac{1}{2}(150 - 90)$$

$$= \frac{1}{2}(60)$$

$$= 30$$

The measure of $\angle LMK$ is **30**.

37. Given:
$$2x^2 - 4x = y + 1$$
$$x + y = 1$$

To solve a linear-quadratic system of equations graphically, graph the equations on the same set of axes. Then find the coordinates of the points at which the graphs intersect.

Graph the quadratic equation:

Solve $2x^2 - 4x = y + 1$ for y, which gives $y = 2x^2 - 4x - 1$. The graph of the quadratic equation $y = 2x^2 - 4x - 1$ is a parabola whose axis of symmetry has an equation of the form $x = -\dfrac{b}{2a}$, where $a = 2$ and $b = -4$:

$$x = -\frac{(-4)}{2(2)}$$

$$= -\left(\frac{-4}{4}\right)$$

$$= 1$$

Hence, the x-coordinate of the vertex of the parabola is 1. To graph $y = 2x^2 - 4x - 1$, follow these steps:

- Step 1: Prepare a table of values that includes three consecutive integer x-values on either side of $x = 1$.

x	$2x^2 - 4x = y$	(x,y)
-2	$2(-2)^2 - 4(-2) - 1 = 8 + 8 - 1 = 15$	$(-2,15)$
-1	$2(-1)^2 - 4(-1) - 1 = 2 + 4 - 1 = 5$	$(-1,5)$
0	$2(0)^2 - 4(0) - 1 = -1$	$(0,-1)$
1	$2(1)^2 - 4(1) - 1 = 2 - 4 - 1 = -3$	$(1,-3)$
2	$2(2)^2 - 4(2) - 1 = 8 - 8 - 1 = -1$	$(2,-1)$
3	$2(3)^2 - 4(3) - 1 = 18 - 12 - 1 = 5$	$(3,5)$
4	$2(4)^2 - 4(4) - 1 = 32 - 16 - 1 = 15$	$(4,15)$

TIP: Use the table feature of your graphing calculator to create a table of values.

- Step 2: Check for symmetry in the y-values listed in the table. Corresponding pairs of points on either side of $x = 1$ have matching y-coordinates. This confirms that $(1,-3)$ is the vertex of the parabola.

- Step 3: Choose a suitable scale. Then plot (–2,15), (–1,5), (0,–1), (1,–3), (2,–1), (3,5), and (4,15) on the grid that is provided. Connect these points with a smooth, U-shaped curve as shown in the accompanying figure. Label the graph with its equation.

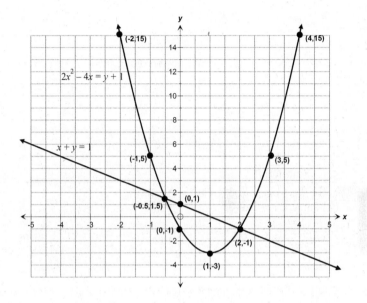

<u>Graph the linear equation on the same set of axes:</u>

To graph $x + y = 1$, find two convenient points on the line. When $x = 0$, $y = 1$. So the line contains the point (0,1). When $x = 2$, $y = 1 - 2 = -1$. So the line also contains the point (2,–1). Plot these two points, and draw a line through them as shown in the accompanying diagram. Label the line with its equation.

State the coordinates of the solution points.

The graphs intersect at **(2,–1)** and **(–0.5,1.5)**, which are the solutions to the given system of equations.

PART IV

38. Given: $\overline{PA}$ and $\overline{PB}$ are tangent to circle O, $\overline{OA}$ and $\overline{OB}$ are radii, $\overline{OP}$ intersects the circle at C. You are required to prove that $\angle AOP \cong \angle BOP$.

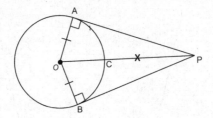

Plan: Prove right triangles OAP and OBP are congruent by $HL \cong HL$.

Statement	Reason
1. $\overline{PA}$ and $\overline{PB}$ are tangent to circle O, $\overline{OA}$ and $\overline{OB}$ are radii.	1. Given.
2. $\overline{OA} \perp \overline{AP}$ and $\overline{OB} \perp \overline{BP}$	2. A radius drawn to a tangent at the point of tangency is perpendicular to the tangent.
3. Angles A and B are right angles.	3. Perpendicular lines intersect to form right angles.
4. Triangles OAP and OBP are right triangles.	4. A triangle that contains a right angle is a right triangle.
5. $\overline{OP} \cong \overline{OP}$ (hypotenuse $\cong$ hypotenuse)	5. Reflexive property of congruence.
6. $\overline{OA} \cong \overline{OB}$ (leg $\cong$ leg)	6. Radii of the same circle are congruent.
7. $\triangle OAP \cong \triangle OBP$	7. Two right triangles are congruent if the hypotenuse and a leg of one right triangle are congruent to the corresponding parts of the other triangle ($HL \cong HL$).
8. $\angle AOP \cong \angle BOP$	8. Corresponding parts of congruent triangles are congruent (CPCTC).

Topic	Question Numbers	Number of Points	Your Points	Your Percentage
1. Logic (negation, conjunction, disjunction, related conditionals, biconditional, logically equivalent statements, logical inference)	29	2		
2. Angle & Line Relationships (vertical angles, midpoint, altitude, median, bisector, supplementary angles, complementary angles, parallel and perpendicular lines)	6	2		
3. Parallel & Perpendicular Planes	8, 18	$2 + 2 = 4$		
4. Angles of a Triangle & Polygon (sum of angles, exterior angle, base angles theorem, angles of a regular polygon)	7, 9, 24	$2 + 2 + 2 = 6$		
5. Triangle Inequalities (side length restrictions, unequal angles opposite unequal sides, exterior angle)	—	—		
6. Trapezoids & Parallelograms (includes properties of rectangle, rhombus, square, and isosceles trapezoid)	11, 16, 25	$2 + 2 + 2 = 6$		
7. Proofs Involving Congruent Triangles	2, 38	$2 + 6 = 8$		
8. Indirect Proof & Mathematical Reasoning	—	—		
9. Ratio & Proportion (includes similar polygons, proportions, proofs involving similar triangles, comparing linear dimensions, and areas of similar triangles)	20, 33	$2 + 2 = 4$		
10. Proportions formed by altitude to hypotenuse of right triangle; Pythagorean theorem	28	2		
11. Coordinate Geometry (slopes of parallel and perpendicular lines; slope-intercept and point-slope equations of a line; applications requiring midpoint, distance, or slope formulas; coordinate proofs)	19, 13, 15, 23, 27, 34	$2 + 2 + 2 + 2 + 2 + 2 = 12$		
12. Solving a Linear-Quadratic System of Equations Graphically	37	4		
13. Transformation Geometry (includes isometries, dilation, symmetry, composite transformations, transformations using coordinates)	3, 22, 26, 32	$2 + 2 + 2 + 2 = 8$		
14. Locus & Constructions (simple and compound locus, locus using coordinates, compass constructions)	1, 21, 30, 35	$2 + 2 + 2 + 4 = 10$		

Topic	Question Numbers	Number of Points	Your Points	Your Percentage
15. Midpoint and Concurrency Theorems (includes joining midpoints of two sides of a triangle, three sides of a triangle, the sides of a quadrilateral, median of a trapezoid; centroid, orthocenter, incenter, and circumcenter)	4	2		
16. Circles and Angle Measurement (includes ≅ tangents from same exterior point, radius ⊥ tangent, tangent circles, common tangents, arcs and chords, diameter ⊥ chord, arc length, area of a sector, center-radius equation, applying transformations)	5, 10, 14, 36	2 + 2 + 2 + 4 = 10		
17. Circles and Similar Triangles (includes proofs, segments of intersecting chords, tangent and secant segments)	17	2		
18. Area of Plane Figures (includes area of a regular polygon)	—	—		
19. Measurement of Solids (volume and lateral area; surface area of a sphere; great circle; comparing similar solids)	12, 31	2 + 2 = 4		

Map to Core Curriculum

Content Band	Item Numbers
Geometric Relationships	8, 12, 18, 31
Constructions	1, 30
Locus	21, 35
Informal and Formal Proofs	2, 4, 5, 6, 7, 11, 16, 17, 19, 20, 24, 25, 28, 29, 33, 36, 38
Transformational Geometry	3, 22, 26, 32
Coordinate Geometry	9, 10, 13, 14, 15, 23, 27, 34, 37

How to Convert Your Raw Score to Your Geometry Regents Examination Score

The conversion chart on the following page must be used to determine your final score on the June 2011 Regents Examination in Geometry. To find your final exam score, locate in the column labeled "Raw Score" the total number of points you scored out of a possible 86 points. Since partial credit is allowed in Parts II, III, and IV of the test, you may need to approximate the credit you would receive for a solution that is not completely correct. Then locate in the adjacent column to the right the scale score that corresponds to your raw score. The scale score is your final Geometry Regents Examination score.

Regents Examination in Geometry—June 2011

Chart for Converting Total Test Raw Scores to Final Examination Scores (Scale Scores)

Raw Score	Scale Score	Raw Score	Scale Score	Raw Score	Scale Score	Raw Score	Scale Score
86	100	64	80	42	66	20	40
85	99	63	80	41	65	19	38
84	97	62	79	40	64	18	37
83	96	61	78	39	63	17	35
82	95	60	78	38	62	16	33
81	94	59	77	37	61	15	32
80	93	58	77	36	60	14	30
79	92	57	76	35	59	13	28
78	91	56	75	34	58	12	26
77	90	55	75	33	57	11	24
76	89	54	74	32	56	10	22
75	88	53	74	31	55	9	20
74	87	52	73	30	54	8	18
73	86	51	72	29	53	7	16
72	86	50	72	28	51	6	13
71	85	49	71	27	50	5	11
70	84	48	70	26	49	4	9
69	83	47	70	25	48	3	7
68	83	46	69	24	46	2	5
67	82	45	68	23	45	1	2
66	81	44	67	22	43	0	0
65	81	43	67	21	42		

Examination
August 2011
Geometry

GEOMETRY REFERENCE SHEET

Volume	Cylinder	$V = Bh$, where B is the area of the base
	Pyramid	$V = \frac{1}{3}Bh$, where B is the area of the base
	Right circular cone	$V = \frac{1}{3}Bh$, where B is the area of the base
	Sphere	$V = \frac{4}{3}\pi r^3$

Lateral area (L)	Right circular cylinder	$L = 2\pi rh$
	Right circular cone	$L = \pi r\ell$, where ℓ is the slant height

Surface area	Sphere	$SA = 4\pi r^2$

PART I

Answer all 28 questions in this part. Each correct answer will receive 2 credits. No partial credit will be allowed. For each question, write in the space provided the number preceding the word or expression that best completes the statement or answers the question.   [56 credits]

1 The statement "x is a multiple of 3, and x is an even integer" is true when x is equal to

(1) 9   (3) 3

(2) 8   (4) 6   1 ____

2 In the diagram below, $\triangle ABC \cong \triangle XYZ$.

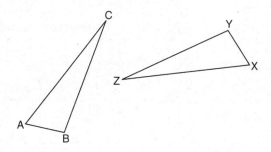

Which statement must be true?

(1) $\angle C \cong \angle Y$ (3) $\overline{AC} \cong \overline{YZ}$

(2) $\angle A \cong \angle X$ (4) $\overline{CB} \cong \overline{XZ}$ 2 ____

3 In the diagram below of $\triangle ABC$, $\overrightarrow{TV} \parallel \overline{BC}$, $AT = 5$, $TB = 7$, and $AV = 10$.

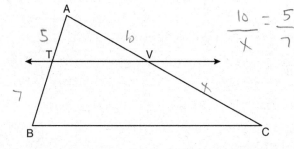

What is the length of $\overline{VC}$?

(1) $3\dfrac{1}{2}$ (3) 14

(2) $7\dfrac{1}{7}$ (4) 24   3 __3__

4 Pentagon $PQRST$ has $\overline{PQ}$ parallel to $\overline{TS}$. After a translation of $T_{2,-5}$, which line segment is parallel to $\overline{P'Q'}$?

(1) $\overline{R'Q'}$ (3) $\overline{T'S'}$

(2) $\overline{R'S'}$ (4) $\overline{T'P'}$ 4 __3__

5 In the diagram below of △*PAO*, $\overline{AP}$ is tangent to circle *O* at point *A*, *OB* = 7, and *BP* = 18.

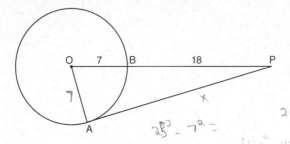

What is the length of $\overline{AP}$?

(1) 10   (3) 17
(2) 12   (4) 24   5 _____

6 A straightedge and compass were used to create the construction below. Arc *EF* was drawn from point *B*, and arcs with equal radii were drawn from *E* and *F*.

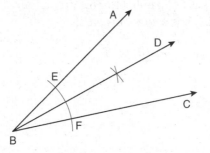

Which statement is *false*?

(1) m∠*ABD* = m∠*DBC*

(2) $\frac{1}{2}$(m∠*ABC*) = m∠*ABD*

(3) 2(m∠*DBC*) = m∠*ABC*

(4) 2(m∠*ABC*) = m∠*CBD*   6 _____

7 What is the length of the line segment whose end-points are $(1,-4)$ and $(9,2)$?

  (1) 5   (3) 10

  (2) $2\sqrt{17}$ (4) $2\sqrt{26}$ 7 __3__

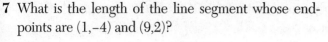

$\sqrt{(2+4)^2 (9-1)^2}$ $\sqrt{128}$

6^2 $36 + 64$

8 What is the image of the point $(2,-3)$ after the transformation $r_{y\text{-axis}}$? $-2,-3$

  (1) $(2,3)$ (3) $(-2,3)$

  (2) $(-2,-3)$ (4) $(-3,2)$ 8 __2__

9 In the diagram below, lines n and m are cut by transversals p and q.

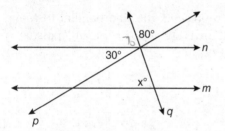

Which value of x would make lines n and m parallel?

  (1) 110   (3) 70

  (2) 80   (4) 50   9 __3__

10 What is an equation of the circle with a radius of 5 and center at $(1,-4)$? $(x-1)^2 + (y+4)^2 = 25$

  (1) $(x + 1)^2 + (y - 4)^2 = 5$

  (2) $(x - 1)^2 + (y + 4)^2 = 5$

  (3) $(x + 1)^2 + (y - 4)^2 = 25$

  (4) $(x - 1)^2 + (y + 4)^2 = 25$ 10 __4__

11 In the diagram below of $\triangle BCD$, side $\overline{DB}$ is extended to point A.

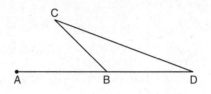

Which statement must be true?

(1) $m\angle C > m\angle D$
(2) $m\angle ABC < m\angle D$
(3) $m\angle ABC > m\angle C$
(4) $m\angle ABC > m\angle C + m\angle D$

11 __3__

12 Which equation represents the line parallel to the line whose equation is $4x + 2y = 14$ and passing through the point $(2,2)$?

$2y = -4x + 14$

(1) $y = -2x$ (3) $y = \frac{1}{2}2x$

$y = -2x + 7$

(2) $y = -2x + 6$ $2 = -4 + 6$ (4) $y = \frac{1}{2}x + 1$

12 __2__

$y = -2x + 6$

13 The coordinates of point A are $(-3a,4b)$. If point A' is the image of point A reflected over the line $y = x$, the coordinates of A' are

(1) $(4b,-3a)$ (3) $(-3a,-4b)$
(2) $(3a,4b)$ (4) $(-4b,-3a)$

13 __1__

$4b, -3a$

14 As shown in the diagram below, $\overline{AC}$ bisects $\angle BAD$ and $\angle B \cong \angle D$.

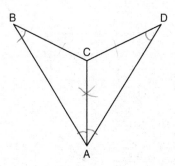

Which method could be used to prove $\triangle ABC \cong \triangle ADC$?

(1) SSS   (3) SAS

(2) AAA   (4) AAS   14 __4__

15 Segment AB is the diameter of circle M. The coordinates of A are $(-4,3)$. The coordinates of M are $(1,5)$. What are the coordinates of B?

(1) (6,7)   (3) (−3,8)

(2) (5,8)   (4) (−5,2)   15 __1__

16 In the diagram below, $\overrightarrow{AB}$ is perpendicular to plane
    AEFG.

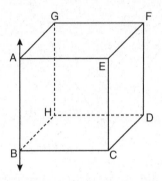

Which plane must be perpendicular to plane *AEFG*?

(1) *ABCE*   (3) *CDFE*

(2) *BCDH*   (4) *HDFG*   16 _____

17 How many points are both 4 units from the origin
    and also 2 units from the line $y = 4$?

(1) 1   (3) 3

(2) 2   (4) 4   17 _____

18 When solved graphically, what is the solution to the following system of equations?

$$y = x^2 - 4x + 6$$
$$y = x + 2$$

(1) (1,4)　　　　　　　　(3) (1,3) and (4,6)

(2) (4,6)　　　　　　　　(4) (3,1) and (6,4)　　　18 _____

19 Triangle PQR has angles in the ratio of 2:3:5. Which type of triangle is $\triangle PQR$?

(1) acute　　　　　　　(3) obtuse

(2) isosceles　　　　　　(4) right　　　　　　19 _____

20 Plane $\mathcal{A}$ is parallel to plane $\mathcal{B}$. Plane C intersects plane $\mathcal{A}$ in line m and intersects plane $\mathcal{B}$ in line n. Lines m and n are

(1) intersecting　　　　(3) perpendicular

(2) parallel　　　　　　(4) skew　　　　　　　20 _____

21 The diagonals of a quadrilateral are congruent but do *not* bisect each other. This quadrilateral is

(1) an isosceles trapezoid

(2) a parallelogram

(3) a rectangle

(4) a rhombus　　　　　　　　　　　　21 _____

22 What is the slope of a line perpendicular to the line represented by the equation $x + 2y = 3$?

(1) -2 (3) $-\dfrac{1}{2}$

(2) 2 (4) $\dfrac{1}{2}$ 22 _____

23 A packing carton in the shape of a triangular prism is shown in the diagram below.

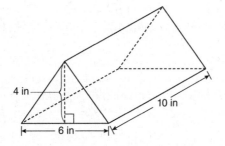

4 in

10 in

6 in

What is the volume, in cubic inches, of this carton?

(1) 20   (3) 120
(2) 60   (4) 240   23 _____

24 In the diagram below of circle O, diameter $\overline{AOB}$ is perpendicular to chord $\overline{CD}$ at point E, $OA = 6$, and $OE = 2$.

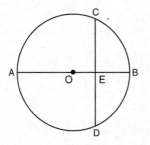

What is the length of $\overline{CE}$?

(1) $4\sqrt{3}$ (3) $8\sqrt{2}$

(2) $2\sqrt{3}$ (4) $4\sqrt{2}$ 24 _____

25 What is the measure of each interior angle of a regular hexagon?

(1) $60°$ (3) $135°$

(2) $120°$ (4) $270°$ 25 _____

26 Which equation represents the perpendicular bisector of $\overline{AB}$ whose endpoints are $A(8,2)$ and $B(0,6)$?

(1) $y = 2x - 4$ (3) $y = -\dfrac{1}{2}x + 6$

(2) $y = -\dfrac{1}{2}x + 2$ (4) $y = 2x - 12$ 26 _____

27 As shown in the diagram below, a kite needs a vertical and a horizontal support bar attached at opposite corners. The upper edges of the kite are 7 inches, the side edges are x inches, and the vertical support bar is $(x + 1)$ inches.

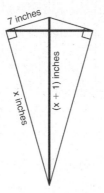

What is the measure, in inches, of the vertical support bar?

(1) 23   (3) 25

(2) 24   (4) 26   27 _____

28 Given three distinct quadrilaterals, a square, a rectangle, and a rhombus, which quadrilaterals must have perpendicular diagonals?

(1) the rhombus, only

(2) the rectangle and the square

(3) the rhombus and the square

(4) the rectangle, the rhombus, and the square   28 _____

PART II

Answer all 6 questions in this part. Each correct answer will receive 2 credits. Clearly indicate the necessary steps, including appropriate formula substitutions, diagrams, graphs, charts, etc. For all questions in this part, a correct numerical answer with no work shown will receive only 1 credit.   [12 credits]

29 In the diagram below, trapezoid $ABCD$, with bases $\overline{AB}$ and $\overline{DC}$, is inscribed in circle O, with diameter $\overline{DC}$. If m $\widehat{AB}$ = 80, find m $\widehat{BC}$.

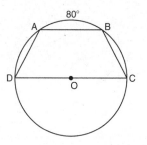

30 On the diagram of $\triangle ABC$ shown below, use a compass and straightedge to construct the perpendicular bisector of $\overline{AC}$. [Leave all construction marks.]

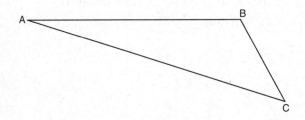

31 A sphere has a diameter of 18 meters. Find the volume of the sphere, in cubic meters, in terms of π.

32 Write an equation of the circle graphed in the diagram below.

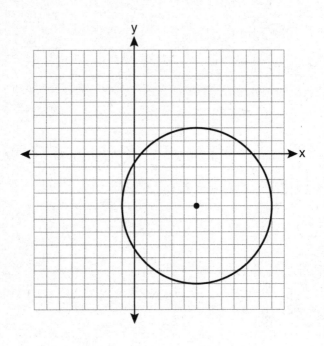

33 The diagram below shows $\triangle ABC$, with $\overline{AEB}$, $\overline{ADC}$, and $\angle ACB \cong \angle AED$. Prove that $\triangle ABC$ is similar to $\triangle ADE$.

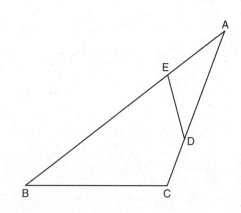

34 Triangle *ABC* has vertices *A*(3,3), *B*(7,9), and *C*(11,3). Determine the point of intersection of the medians, and state its coordinates. [The use of the set of axes below is optional.]

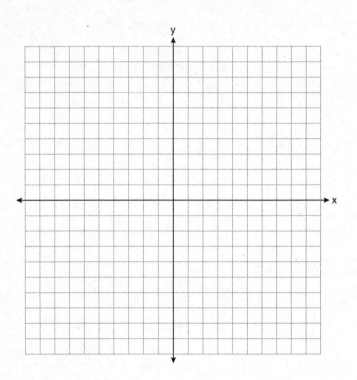

PART III

Answer all 3 questions in this part. Each correct answer will receive 4 credits. Clearly indicate the necessary steps, including appropriate formula substitutions, diagrams, graphs, charts, etc. For all questions in this part, a correct numerical answer with no work shown will receive only 1 credit. [12 credits]

35 In the diagram below of $\triangle GJK$, H is a point on $\overline{GJ}$, $\overline{HJ} \cong \overline{JK}$, m$\angle G$ = 28, and m$\angle GJK$ = 70. Determine whether $\triangle GHK$ is an isosceles triangle and justify your answer.

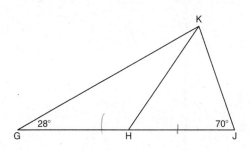

36 As shown on the set of axes below, $\triangle GHS$ has vertices $G(3,1)$, $H(5,3)$, and $S(1,4)$. Graph and state the coordinates of $\triangle G''H''S''$, the image of $\triangle GHS$ after the transformation $T_{-3,1} \circ D_2$.

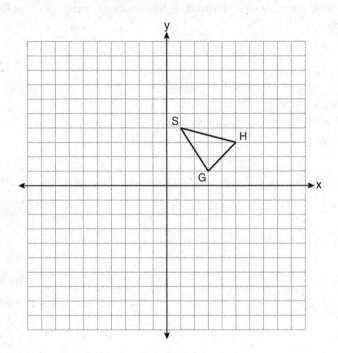

37 In the diagram below, $\triangle ABC \sim \triangle DEF$, $DE = 4$, $AB = x$, $AC = x + 2$, and $DF = x + 6$. Determine the length of $\overline{AB}$. [Only an algebraic solution can receive full credit.]

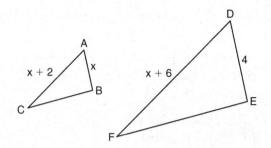

PART IV

Answer the question in this part. A correct answer will receive 6 credits. Clearly indicate the necessary steps, including appropriate formula substitutions, diagrams, graphs, charts, etc. A correct numerical answer with no work shown will receive only 1 credit.   [6 credits]

38 Given: $\triangle ABC$ with vertices $A(-6,-2)$, $B(2,8)$, and $C(6,-2)$ $\overline{AB}$ has midpoint D, $\overline{BC}$ has midpoint E, and $\overline{AC}$ has midpoint F

Prove: $ADEF$ is a parallelogram
        $ADEF$ is not a rhombus

[The use of the grid below is optional.]

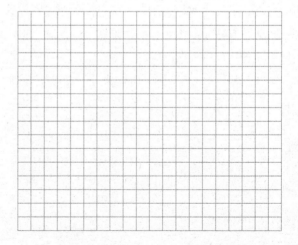

Answers
August 2011
Geometry

Answer Key

PART I

1. (4)	**8.** (2)	**15.** (1)	**22.** (2)
2. (2)	**9.** (3)	**16.** (1)	**23.** (3)
3. (3)	**10.** (4)	**17.** (2)	**24.** (4)
4. (3)	**11.** (3)	**18.** (3)	**25.** (2)
5. (4)	**12.** (2)	**19.** (4)	**26.** (1)
6. (4)	**13.** (1)	**20.** (2)	**27.** (3)
7. (3)	**14.** (4)	**21.** (1)	**28.** (3)

PART II

29. 50

30. See the *Answers Explained* section.

31. 972π

32. $(x - 5)^2 + (y + 4)^2 = 36$

33. See the *Answers Explained* section.

34. (7,5)

PART III

35. It is not an isosceles triangle.

36. $G''(3,3)$, $H''(7,7)$, and $S''(-1,9)$

37. 2

PART IV

38. See the *Answers Explained* section.

In **PARTS II–IV** you are required to show how you arrived at your answers. For sample methods of solutions, see the *Answers Explained* section.

Answers Explained

PART I

1. The given statement is "x is a multiple of 3, and x is an even integer." The statement is true when both conditions are true. Since x is required to be an even integer, you can eliminate choices (1) and (3). Of the remaining two choices, only 6 is a multiple of 3 since $2 \times 3 = 6$.

The correct choice is **(4)**.

2. In the accompanying diagram, $\triangle ABC \cong \triangle XYZ$.

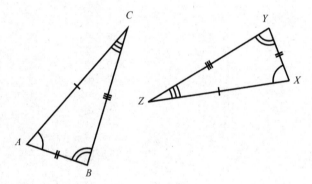

In the congruence statement $\triangle ABC \cong \triangle XYZ$, corresponding vertices have corresponding positions. For example, since A and X occupy the first positions in the congruence statement, they are corresponding vertices so that $\angle A \cong \angle X$. Use the congruence statement to mark off all pairs of corresponding congruent parts as shown in the accompanying diagram. Based on the diagram, evaluate each answer choice in turn until you find the answer choice that contains a pair of congruent parts, as in $\angle A \cong \angle X$.

The correct choice is **(2)**.

3. In the accompanying diagram of $\triangle ABC$, $\overline{TV} \parallel \overline{BC}$, $AT = 5$, $TB = 7$, and $AV = 10$.

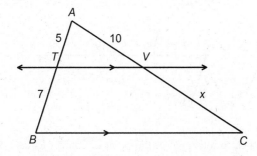

A line that intersects two sides of a triangle and is parallel to the third side divides the sides it intersects proportionately. If x represents the length of $\overline{VC}$:

$$\frac{5}{7} = \frac{10}{x}$$
$$5x = 70$$
$$\frac{5x}{5} = \frac{70}{5}$$
$$x = 14$$

The correct choice is **(3)**.

4. A pentagon $PQRST$ has $\overline{PQ}$ parallel to $\overline{TS}$. You are asked to determine which line segment is parallel to $\overline{P'Q'}$ after a translation of $T_{2,-5}$. Since the translation $T_{2,-5}$ simply slides the pentagon in the horizontal and vertical directions while preserving angle measure, the images of $\overline{PQ}$ and $\overline{TS}$ will also be parallel. Thus, $\overline{P'Q'} \parallel \overline{T'S'}$.

The correct choice is **(3)**.

5. In the accompanying diagram of $\triangle PAO$, $\overline{AP}$ is tangent to circle O at point A, $OB = 7$, and $BP = 18$. You are asked to find the length of $\overline{AP}$.

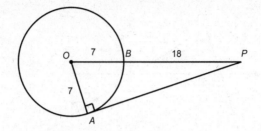

- A radius drawn to a tangent at the point of contact forms a right angle with the tangent. Hence, $\angle OAP$ is a right angle. In right triangle OAP, OP is the hypotenuse where $OP = 7 + 18 = 25$.
- Since $\overline{OA}$ is a radius, $OA = OB = 7$.
- Use the Pythagorean theorem to find $\overline{AP}$:

$$(7)^2 + (AP)^2 = (25)^2$$

$$49 + (AP)^2 = 625$$

$$(AP)^2 = 625 - 49$$

$$\sqrt{(AP)^2} = \sqrt{576}$$

$$AP = 24$$

The correct choice is **(4)**.

6. The accompanying diagram shows the construction of $\overrightarrow{BD}$, the bisector of $\angle ABC$.

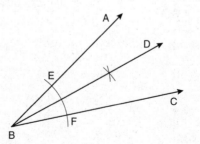

To help determine which statement among the answer choices is false, suppose that m$\angle ABC$ = 60. Then m$\angle ABD$ = m$\angle DBC$ = 30. It is easy to see using this example that choices (1), (2), and (3) contain true statements. The statement in choice (4), however, is false since 2(60) ≠ 30.

The correct choice is (**4**).

7. To find the length, d, of the line segment with endpoints at (1,–4) and (9,2), use the distance formula:

$$d = \sqrt{(\Delta x)^2 + (\Delta y)^2}$$

In this formula, Δx represents the difference in the x-coordinates of the two given points and Δy represents the difference in their y-coordinates, with the subtractions performed in the same order:

$$d = \sqrt{(2-(-4))^2 + (9-1)^2}$$
$$= \sqrt{(2+4)^2 + (8)^2}$$
$$= \sqrt{36+64}$$
$$= \sqrt{100}$$
$$= 10$$

The correct choice is (**3**).

8. The notation $r_{y\text{-}axis}$ represents a reflection over the y-axis. If a point is reflected over the y-axis, its image lies on the opposite side of the y-axis and the same distance from that axis. In general, $r_{y\text{-}axis}(x,y) = (-x,y)$. So the image of (2,–3) after the transformation $r_{y\text{-}axis}$ is (–2,–3).

The correct choice is (**2**).

9. In the accompanying diagram, lines n and m are cut by transversals p and q.

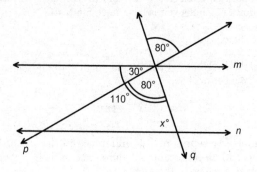

- Since vertical angles are congruent, the angle opposite the 80° angle also measures 80°.

- As shown in the accompanying diagram, an interior angle on the same side of the transversal as angle x measures 30° + 80° or 110°.

- Two lines are parallel if consecutive angles on the same side of the transversal are supplementary. Thus, lines m and n are parallel if $x + 110 = 180$ or $x = 180 - 110 = 70$.

The correct choice is **(3)**.

10. The center-radius form of the equation of a circle is $(x - h)^2 + (y - k)^2 = r^2$ where (h,k) is the center of the circle and r is the radius. The radius of a given circle is 5 and the center is at $(1,-4)$. Write an equation of this circle in center-radius form by replacing r with 5, h with 1, and k with -4:

$$\left(x - \underset{h}{1}\right)^2 + \left(y - \underset{k}{(-4)}\right)^2 = \underset{r}{5^2}$$

$$(x-1)^2 + (y+4)^2 = 25$$

The correct choice is **(4)**.

11. In the accompanying diagram of $\triangle BCD$, $\angle ABC$ is an exterior angle of the triangle. So $\angle ABC$ must be greater in measure than the measures of either of the two non-adjacent interior angles (angles C or D). Thus, $m\angle ABC > m\angle C$, which corresponds to choice (3).

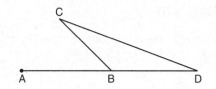

The correct choice is **(3)**.

12. To find an equation that represents the line parallel to the line $4x + 2y = 14$ and passing through the point $(2,2)$, rewrite the given equation in slope-intercept form $y = mx + b$ where m is the slope of the line and b is its y-intercept:

- Dividing each member of the equation $4x + 2y = 14$ by 2 gives $2x + y = 7$. Rearranging the equation gives $y = -2x + 7$. The slope of the given line is -2, the coefficient of the x-term.

- Because parallel lines have the same slope, the slope of the required line is also -2. Its equation has the form $y = -2x + b$.

- The coordinates of the point $(2,2)$ must satisfy the equation of the line. Find b by substituting 2 for y and 2 for x in $y = -2x + b$:

$$2 = -2(2) + b$$

$$2 = -4 + b$$

$$6 = b$$

- Since $m = -2$ and $b = 6$, an equation of the required line is $y = -2x + 6$.

The correct choice is **(2)**.

13. The image of any point (x,y) when reflected over the line $y = x$ is the point (y,x). To find the coordinates of the image of point $A(-3a,4b)$ when reflected over the line $y = x$, simply interchange the x- and y-coordinates which gives $A'(4b,-3a)$.

The correct choice is **(1)**.

14. It is given that in the accompanying diagram, $\overline{AC}$ bisects $\angle BAD$ and $\angle B \cong \angle D$.

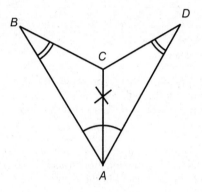

Mark off the diagram with the information given. Since $\overline{AC} \cong \overline{AC}$, two angles and a side opposite one of these angles in $\triangle ABC$ are congruent to the corresponding parts of $\triangle ADC$. Hence, the AAS method could be used to prove $\triangle ABC \cong \triangle ADC$.

The correct choice is **(4)**.

15. It is given that $\overline{AB}$ is the diameter of circle M, the coordinates of A are $(-4,3)$ and the coordinates of M are $(1,5)$. You are asked to find the coordinates of $B(x,y)$. Since M is the center of the circle, $M(1,5)$ is the midpoint of $\overline{AB}$. So the coordinates of point M are the averages of the corresponding coordinates of points A and B.

$$1 = \frac{-4+x}{2} \qquad \text{and} \qquad 5 = \frac{3+y}{2}$$

$$2 = -4 + x \qquad\qquad\qquad 10 = 3 + y$$

$$6 = x \qquad\qquad\qquad\qquad 7 = y$$

The coordinates of B are $(6,7)$.

The correct choice is **(1)**.

16. In the accompanying diagram, $\overrightarrow{AB}$ is perpendicular to plane *AEFG*.

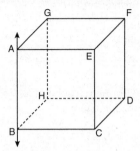

Since $\overrightarrow{AB}$ lies in plane *ABCE*, plane *ABCE* must also be perpendicular to plane *AEFG*.

The correct choice is (**1**).

17. To find the number of points that are 4 units from the origin and also 2 units from the line $y = 4$, sketch the two locus conditions on the same set of axes.

- The locus of points 4 units from the origin is a circle whose center is at the origin and that has a radius of 4 as shown in the accompanying diagram.

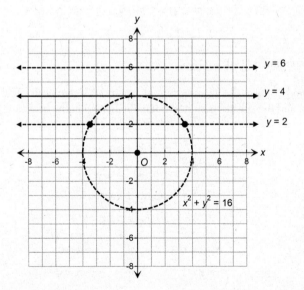

- The locus of points 2 units from the line $y = 4$ are two parallel lines on either side of this line and each 2 units from it as shown by the broken lines in the accompanying diagram.

- The two locus conditions intersect in exactly two points where the line $y = 2$ intersects the circle.

The correct choice is **(2)**.

18. You are asked to determine from among the answer choices the solution of the following system of equations:

$$y = x^2 - 4x + 6$$
$$y = x + 2$$

<u>Method 1</u>: Substitute the answer choices into the equations. Examine the answer choices and eliminate any choice that does not satisfy the linear equation, as in choices (1) and (4).

The two remaining answer choices, (2) and (3), both include the ordered pair (4,6). So that ordered pair must belong to the solution set. Choice (3) includes (1,3) while choice (2) does not. Determine whether (1,3) satisfies both equations by setting $x = 1$ and $y = 3$:

$$3 \stackrel{?}{=} 3^2 - 4(3) + 6 = 9 - 12 + 6 = 3 \quad ✔$$
$$3 \stackrel{?}{=} 1 + 2 = 3 \quad ✔$$

The solution is (1,3) and (4,6).

<u>Method 2</u>: Solve the given system of equations algebraically. Use the linear equation and the substitution method to eliminate y in the quadratic equation:

$$x^2 - 4x + 6 = \overbrace{x + 2}^{y}$$
$$x^2 - 4x - x + 6 - 2 = 0$$
$$x^2 - 5x + 4 = 0$$
$$(x-1)(x-4) = 0$$

$$x - 1 = 0 \qquad \text{or} \qquad x - 4 = 0$$
$$x = 1 \qquad \text{or} \qquad x = 4$$

Find the corresponding values of y by substituting the solution values of x into the linear equation.

- If $x = 1$, then $y = 1 + 2 = 3$. So $(1,3)$ is a solution.

- If $x = 4$, then $y = 4 + 2 = 6$. So $(4,6)$ is a solution.

<u>Method 3</u>: Solve graphically.

Using your graphing calculator, graph the two equations on the same set of axes using a friendly window such as $[-9.4,9.4] \times [-9.3,9.3]$. Then use the TRACE feature to find the coordinates of the points at which the two graphs intersect as shown in the accompanying diagram.

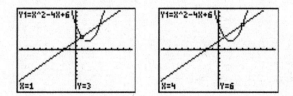

The correct choice is **(3)**.

19. If the measures of the angles of $\triangle PQR$ are in the ratio of 2:3:5, then $2x + 3x + 5x = 180$. So $10x = 180$, and $x = 18$. Hence, the three angles of the triangle measure:

$$2x = 2(18) = 36$$

$$3x = 3(18) = 54$$

$$5x = 5(18) = 90$$

Since the triangle contains a 90-degree angle, $\triangle PQR$ is a right triangle.

The correct choice is **(4)**.

20. Plane $\mathcal{A}$ is parallel to plane $\mathcal{B}$. Plane C intersects plane $\mathcal{A}$ in line m and intersects plane $\mathcal{B}$ in line n. You are asked to determine the relationship between lines m and n.

If a plane intersects two parallel planes, the intersections form two parallel lines as illustrated in the accompanying figure.

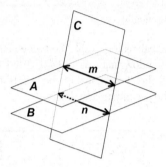

The correct choice is **(2)**.

21. You are asked to determine from among the answer choices the quadrilateral in which the diagonals are congruent but do *not* bisect each other. The diagonals of a parallelogram bisect each other. Since a rectangle and a rhombus are parallelograms, choices (2), (3), and (4) can be eliminated.

The correct choice is **(1)**.

22. To find the slope of a line that is perpendicular to the line whose equation is $x + 2y = 3$, first write the given equation in the slope-intercept form $y = mx + b$ where m is the slope of the line and b is its y-intercept:

$$x + 2y = 3$$

$$2y = -x + 3$$

$$\frac{2y}{2} = \frac{-x}{2} + \frac{3}{2}$$

$$y = -\frac{1}{2}x + \frac{3}{2}$$

- The slope of the given line is $-\frac{1}{2}$, the coefficient of x.

- Perpendicular lines have slopes that are negative reciprocals. The negative reciprocal of $-\frac{1}{2}$ is 2 since $\left(-\frac{1}{2}\right) \times (2) = -1$.

- A line with a slope of 2 is perpendicular to the given line.

The correct choice is **(2)**.

23. You are asked to determine the volume of a packing carton that is in the shape of a triangular prism as shown in the accompanying diagram.

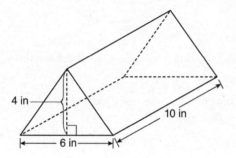

The volume of a prism is the area of its base times its height. The height of the prism is the distance between the two bases.

The area of a triangle is one-half its base times its height. Since the base of the prism is a triangle, the area of its base is $\frac{1}{2}(6)(4) = 12$ in^2.

The distance between the triangular bases is given as 10 inches. So the volume of the prism is 12 in$^2 \times$ 10 in = 120 in^3.

The correct choice is **(3)**.

24. In the accompanying diagram of circle O, diameter $\overline{AOB}$ is perpendicular to chord $\overline{CD}$ at point E, $OA = 6$, and $OE = 2$. You are asked to find the length of $\overline{CE}$.

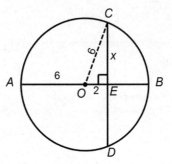

- Draw radius $\overline{OC}$. Since all radii of a circle have the same length, $OC = OA = 6$.

- In right triangle OEC, use the Pythagorean theorem to find x where x represents the length of $\overline{CE}$:

$$x^2 + 2^2 = 6^2$$

$$x^2 + 4 = 36$$

$$x^2 + 4 - 4 = 36 - 4$$

$$x^2 = 32$$

$$x = \sqrt{32}$$

$$= \sqrt{16} \cdot \sqrt{2}$$

$$= 4\sqrt{2}$$

The correct choice is (**4**).

25. The measure of each interior angle of a regular polygon with n sides is given by the expression $\dfrac{(n-2)180°}{n}$ where the numerator represents the sum of the measures of the interior angles of any n-sided polygon. To find the measure of each interior angle of a regular hexagon, evaluate the expression for $n = 6$:

$$\frac{(6-2)180°}{6} = \frac{\overset{30°}{4(\cancel{180°})}}{\cancel{6}} = 120°$$

The correct choice is (**2**).

26. You are asked to identify the equation that represents the perpendicular bisector of $\overline{AB}$ whose endpoints are $A(8,2)$ and $B(0,6)$. The perpendicular bisector of $\overline{AB}$ passes through the midpoint of $\overline{AB}$ and has a slope that is the negative reciprocal of the slope of $\overline{AB}$.

- The coordinates of the midpoint of $\overline{AB}$ are the averages of the corresponding coordinates of its endpoints. So the coordinates of the midpoint of $\overline{AB}$ are $\left(\dfrac{8+0}{2}, \dfrac{2+6}{2}\right) = (4,4)$.

- Use the slope formula to find the slope, m, of $\overline{AB}$:

$$m = \frac{\Delta y}{\Delta x}$$
$$= \frac{6-2}{0-8}$$
$$= \frac{4}{-8}$$
$$= -\frac{1}{2}$$

Since the slopes of perpendicular lines are negative reciprocals, the slope of the perpendicular bisector of $\overline{AB}$ is 2, the negative reciprocal of $-\dfrac{1}{2}$.

• An equation of the perpendicular bisector of $\overline{AB}$ has the form $y = mx + b$ where $m = 2$. The value of b can be determined by substituting the coordinates of the midpoint of $\overline{AB}$ into the equation $y = 2x + b$. Setting $x = 4$ and $y = 4$ gives:

$$4 = 2(4) + b$$
$$4 = 8 + b$$
$$-4 = b$$

Since $m = 2$ and $b = -4$, the equation $y = 2x - 4$ represents the perpendicular bisector of $\overline{AB}$.

The correct choice is **(1)**.

27. You are asked to determine the length of the vertical support bar marked as having a length of $(x + 1)$ inches in the accompanying diagram.

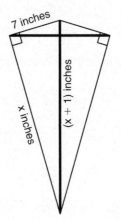

The vertical support bar forms the hypotenuse of a right triangle whose legs have lengths represented by x and 7. Use the Pythagorean theorem to find x:

$$x^2 + 7^2 = (x+1)^2$$

$$x^2 + 49 = x^2 + 2x + 1$$

$$49 - 1 = 2x + 1 - 1$$

$$\frac{48}{2} = \frac{2x}{2}$$

$$24 = x$$

The length of the vertical support bar is $x + 1 = 24 + 1 = 25$ inches.

The correct choice is **(3)**.

28. Of the three distinct quadrilaterals—square, rectangle, and rhombus—only a rhombus and a square have perpendicular diagonals. The diagonals of a rectangle are congruent.

The correct choice is **(3)**.

PART II

29. In the accompanying diagram, trapezoid *ABCD*, with bases $\overline{AB}$ and $\overline{DC}$, is inscribed in circle *O* with diameter $\overline{DC}$. You are asked to find m $\overparen{BC}$ given that m $\overparen{AB}$ = 80.

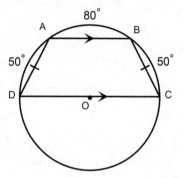

Because a semicircle measures 180°, m $\overparen{AD}$ + 80 + m $\overparen{BC}$ =180. So m $\overparen{AD}$ + m $\overparen{BC}$ =100.

Since the bases of a trapezoid are parallel, chords $\overline{AB}$ and $\overline{DC}$ cut off arcs that are equal in measure. This means m $\overparen{AD}$ = m $\overparen{BC}$ =50 as indicated in the accompanying figure.

So, m $\overparen{BC}$ = **50**.

30. Using a compass, straightedge, and the accompanying diagram of $\triangle ABC$, you must construct the perpendicular bisector of side $\overline{AC}$.

- Set the radius length of your compass to any convenient length that is more than one-half the length of $\overline{AC}$.

- Use this compass setting and points *A* and *C* as centers to draw arcs above and below $\overline{AC}$. Label the points of intersection of the arcs as *P* and *Q*. This is shown in the accompanying diagram.

- Use a straightedge to draw $\overrightarrow{PQ}$, which is the perpendicular bisector of $\overline{AC}$.

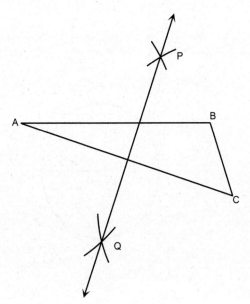

31. To find the volume, in cubic meters, of a sphere whose diameter is 18 meters, use the formula $V = \frac{4}{3}\pi r^3$ where r is the radius of a sphere whose volume is V. This formula is provided on the reference sheet at the beginning of this exam. Since the diameter of the sphere is 18 meters, its radius is 9 meters:

$$V = \frac{4}{3}\pi(9^3)$$

$$= \frac{4}{3}\pi(729)$$

$$= 972\pi$$

The volume of the cube in terms of π is **972π cubic meters**.

32. You are asked to write an equation of the circle graphed in the accompanying diagram.

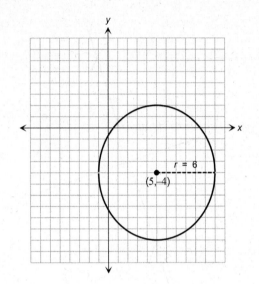

Count the unit boxes to find that the center of the circle is at (5,–4) and a radius of the circle is 6. Use the center-radius form of the equation of a circle $(x - h)^2 + (y - k)^2 = r^2$. Replace h with 5, k with –4, and r with 6:

$$(x - 5)^2 + (y - (-4))^2 = 6^2$$

The equation of the circle is $(x - 5)^2 + (y + 4)^2 = 36$.

33. The accompanying diagram shows $\triangle ABC$ with $\overline{AEB}$, $\overline{ADC}$, and $\angle ACB \cong \angle AED$.

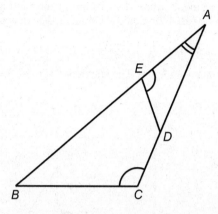

Plan: To prove that $\triangle ABC$ is similar to $\triangle ADE$, show that two angles of one triangle are congruent to the corresponding angles of the other triangle (AA).

Statement	Reason
1. $\triangle ABC$ with $\overline{AEB}$, $\overline{ADC}$, and $\angle ACB \cong \angle AED$	1. Given.
2. $\angle A \cong \angle A$	2. Reflexive property of congruence.
3. $\triangle ABC \sim \triangle ADE$	3. Two triangles are similar if two angles of one triangle are congruent to the corresponding angles of the other triangle (AA).

34. Triangle ABC has vertices $A(3,3)$, $B(7,9)$, and $C(11,3)$. You must determine the point of intersection of the medians and state its coordinates. To do this, draw $\triangle ABC$ on the grid that is provided as shown in the accompanying diagram. The three medians of a triangle meet at a point that divides each median into two segments whose lengths are in the ratio of 1:2. Locate this point, called the centroid of the triangle, on the median drawn from B to $\overline{AC}$.

The midpoint of $\overline{AC}$ is $\left(\dfrac{3+11}{2}, \dfrac{3+3}{2}\right) = M(7,3)$. Draw median $\overline{BM}$ as illustrated in the accompanying figure. The length of $\overline{BM}$ is $9 - 3 = 6$.

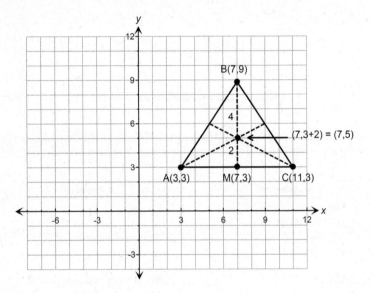

The point of intersection of the three medians of $\triangle ABC$ divides $\overline{BM}$ into two segments whose lengths are in the ratio of 1:2. Since $BM = 6$, the length of the shorter of the two segments is 2 and the length of the longer segment is 4. The longer of these two segments is the one located closer to B, the vertex of the triangle from which the median is drawn as shown in the accompanying diagram.

So the point of intersection of the medians of $\triangle ABC$ is located 2 units directly above $M(7,3)$ at $(7,5)$.

The coordinates of the point of intersection of the medians is **(7,5)**.

PART III

35. In the accompanying diagram of $\triangle GJK$, H is a point on $\overline{GJ}$, $\overline{HJ} \cong \overline{JK}$, $m\angle G = 28$, and $m\angle GJK = 70$.

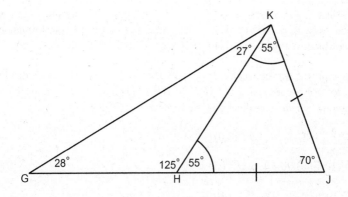

To decide whether $\triangle GHK$ is an isosceles triangle, determine if two of its angles are congruent. Find $m\angle GHK$ and $m\angle GKH$.

- In $\triangle HJK$, $m\angle KHJ + m\angle JKH + 70 = 180$ or $m\angle KHJ + m\angle JKH = 110$. Since

  $\overline{HJ} \cong \overline{JK}$, $m\angle KHJ = m\angle JKH$. So $m\angle KHJ = m\angle JKH = \frac{1}{2}(110) = 55$.

- Angles GHK and KHJ are supplementary. So $m\angle GHK + 55 = 180$ and $m\angle GHK = 180 - 55 = 125$.

- In $\triangle GHK$, $28 + 125 + m\angle GKH = 180$. So $m\angle GKH = 180 - 153 = 27$.

Since $\triangle GHK$ does not have two congruent angles, **it is *not* an isosceles triangle**.

36. As shown on the accompanying set of coordinate axes, $\triangle GHS$ has vertices $G(3,1)$, $H(5,3)$, and $S(1,4)$. You are asked to graph and state the coordinates of $\triangle G''H''S''$, the image of $\triangle GHS$ after the transformation $T_{-3,1} \circ D_2$. The notation $T_{-3,1} \circ D_2$ means a dilation with a scale factor of 2 followed by the translation of $T_{-3,1}$.

To dilate $\triangle GHS$ using a scale factor of 2, multiply the coordinates of each of its vertices by 2. If $\triangle G'H'S'$ is the image of $\triangle GHS$ after the transformation D_2, the coordinates of the vertices of $\triangle G'H'S'$ are $G'(6,2)$, $H'(10,6)$, and $S'(2,8)$.

To apply the transformation $T_{-3,1}$ to $\triangle G'H'S'$, add –3 to each x-coordinate and add 1 to each y-coordinate. If $\triangle G''H''S''$ is the image of $\triangle G'H'S'$ after the transformation $T_{-3,1}$, the coordinates of the vertices of $\triangle G''H''S''$ are:

$G''(6 + (-3), 2 + 1) = \textbf{\textit{G}''(3,3)}$

$H''(10 + (-3), 6 + 1) = \textbf{\textit{H}''(7,7)}$

$S''(2 + (-3), 8 + 1) = \textbf{\textit{S}''(-1,9)}$

Graph $\triangle G''H''S''$ as shown in the accompanying figure.

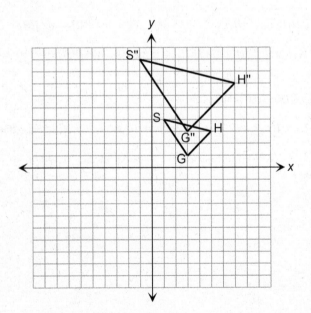

37. In the accompanying diagram, $\triangle ABC \sim \triangle DEF$, $DE = 4$, $AB = x$, $AC = x + 2$, and $DF = x + 6$. You need to determine the length of $\overline{AB}$.

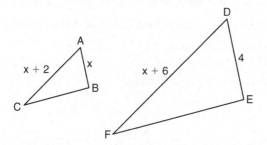

Since the lengths of corresponding sides of similar triangles are in proportion:

$$\frac{AC}{DF} = \frac{AB}{DE}$$

$$\frac{x+2}{x+6} = \frac{x}{4}$$

$$x(x+6) = 4(x+2)$$

$$x^2 + 6x = 4x + 8$$

$$x^2 + 6x - 4x - 8 = 0$$

$$x^2 + 2x - 8 = 0$$

$$(x-2)(x+4) = 0$$

$$x - 2 = 0 \quad \text{or} \quad x + 4 = 0$$

$$x = 2 \qquad\qquad x = -4 \leftarrow \text{reject since } x$$
$$\text{must be positive}$$

Thus $AB = x = 2$.

The length of $\overline{AB}$ is **2**.

PART IV

38. Given: $\triangle ABC$ with vertices $A(-6,-2)$, $B(2,8)$, and $G(6,-2)$.

$\overline{AB}$ has midpoint D, $\overline{BC}$ has midpoint E, and $\overline{AC}$ has midpoint F.

Prove: $ADEF$ is a parallelogram
          $ADEF$ is *not* a rhombus

Sketch the triangle on the grid provided.

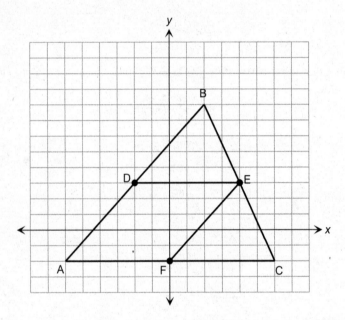

Use the midpoint formula to locate the midpoints of each side of the triangle:

$$D\left(\frac{-6+2}{2},\frac{-2+8}{2}\right) = D(-2,3)$$

$$E\left(\frac{2+6}{2},\frac{8+(-2)}{2}\right) = E(4,3)$$

$$F\left(\frac{-6+6}{2},\frac{-2+(-2)}{2}\right) = F(0,-2)$$

To prove *ADEF* is a parallelogram, show that its diagonals have the same midpoint and, as a result, bisect each other.

Midpoint of diagonal $\overline{AE}$ = $\left(\dfrac{-6+4}{2},\dfrac{-2+3}{2}\right)$ = $\left(-1,\dfrac{1}{2}\right)$

Midpoint of diagonal $\overline{FD}$ = $\left(\dfrac{0+-2}{2},\dfrac{-2+3}{2}\right)$ = $\left(-1,\dfrac{1}{2}\right)$

Since the diagonals have the same midpoint, the diagonals of quadrilateral *ADEF* bisect each other. If the diagonals of a quadrilateral bisect each other, the quadrilateral is a parallelogram. So *ADEF* is a parallelogram.

Note: You could also prove that *ADEF* is a parallelogram by one of these other methods. Use the distance formula to show that opposite sides are equal in length. Use the slope formula to show that opposites sides are parallel. Show that one pair of sides are both parallel and equal in length.

A rhombus is an equilateral parallelogram. So adjacent sides must have the same length. To prove *ADEF* is not a rhombus, show that a pair of adjacent sides do *not* have the same length.

Find the length of $\overline{AF}$. Since $\overline{AF}$ is a horizontal line segment, its length is the positive difference in the *x*-coordinates of its endpoints:

$$AF = 0 - (-6) = 6$$

Use the distance formula to find the length of $\overline{AD}$:

$$\begin{aligned}
AD &= \sqrt{\left(-2-(-6)\right)^2 + \left(3-(-2)\right)^2} \\
&= \sqrt{\left(-2+6\right)^2 + \left(3+2\right)^2} \\
&= \sqrt{4^2 + 5^2} \\
&= \sqrt{16+25} \\
&= \sqrt{41}
\end{aligned}$$

Since $\overline{AF} \neq \overline{AD}$, parallelogram *ADEF* is *not* a rhombus.

Topic	Question Numbers	Number of Points	Your Points	Your Percentage
1. Logic (negation, conjunction, disjunction, related conditionals, biconditional, logically equivalent statements, logical inference)	1	2		
2. Angle & Line Relationships (vertical angles, midpoint, altitude, median, bisector, supplementary angles, complementary angles, parallel and perpendicular lines)	9	2		
3. Parallel & Perpendicular Planes	16, 20	2 + 2 = 4		
4. Angles of a Triangle & Polygon (sum of angles, exterior angle, base angles theorem, angles of a regular polygon)	11, 19, 25, 35	2 + 2 + 2 + 4 = 10		
5. Triangle Inequalities (side length restrictions, unequal angles opposite unequal sides, exterior angle)	—	—		
6. Trapezoids & Parallelograms (includes properties of rectangle, rhombus, square, and isosceles trapezoid)	21, 28	2 + 2 = 4		
7. Proofs Involving Congruent Triangles	2, 14	2 + 2 = 4		
8. Indirect Proof & Mathematical Reasoning	—	—		
9. Ratio & Proportion (includes similar polygons, proportions, proofs involving similar triangles, comparing linear dimensions, and areas of similar triangles)	3, 33, 37	2 + 2 + 4 = 8		
10. Proportions formed by altitude to hypotenuse of right triangle; Pythagorean theorem	24, 27	2 + 2 = 4		
11. Coordinate Geometry (slopes of parallel and perpendicular lines; slope-intercept and point-slope equations of a line; applications requiring midpoint, distance, or slope formulas; coordinate proofs)	7, 12, 15, 22, 26, 34, 38	2 + 2 + 2 + 2 + 2 + 2 + 6 = 18		
12. Solving a Linear-Quadratic System of Equations Graphically	18	2		
13. Transformation Geometry (includes isometries, dilation, symmetry, composite transformations, transformations using coordinates)	4, 8, 13, 36	2 + 2 + 2 + 4 = 10		
14. Locus & Constructions (simple and compound locus, locus using coordinates, compass constructions)	6, 17, 30	2 + 2 + 2 = 6		

Topic	Question Numbers	Number of Points	Your Points	Your Percentage
15. Midpoint and Concurrency Theorems (includes joining midpoints of two sides of a triangle, three sides of a triangle, the sides of a quadrilateral, median of a trapezoid; centroid, orthocenter, incenter, and circumcenter)	—	—		
16. Circles and Angle Measurement (includes ≅ tangents from same exterior point, radius ⊥ tangent, tangent circles, common tangents, arcs and chords, diameter ⊥ chord, arc length, area of a sector, center-radius equation, applying transformations)	5, 10, 29, 32	$2 + 2 + 2 + 2 = 8$		
17. Circles and Similar Triangles (includes proofs, segments of intersecting chords, tangent and secant segments)	—	—		
18. Area of Plane Figures (includes area of a regular polygon)	—	—		
19. Measurement of Solids (volume and lateral area; surface area of a sphere; great circle; comparing similar solids)	23, 31	$2 + 2 = 4$		

Map to Core Curriculum

Content Band	Item Numbers
Geometric Relationships	16, 20, 23, 31
Constructions	6, 30
Locus	17, 34
Informal and Formal Proofs	1, 2, 3, 5, 9, 11, 14, 19, 21, 24, 25, 27, 28, 29, 33, 35, 37
Transformational Geometry	4, 8, 13, 36
Coordinate Geometry	7, 10, 12, 15, 18, 22, 26, 32, 38

How to Convert Your Raw Score to Your Geometry Regents Examination Score

The conversion chart on the following page must be used to determine your final score on the August 2011 Regents Examination in Geometry. To find your final exam score, locate in the column labeled "Raw Score" the total number of points you scored out of a possible 86 points. Since partial credit is allowed in Parts II, III, and IV of the test, you may need to approximate the credit you would receive for a solution that is not completely correct. Then locate in the adjacent column to the right the scale score that corresponds to your raw score. The scale score is your final Geometry Regents Examination score.

Regents Examination in Geometry—August 2011

Chart for Converting Total Test Raw Scores to Final Examination Scores (Scale Scores)

Raw Score	Scale Score	Raw Score	Scale Score	Raw Score	Scale Score	Raw Score	Scale Score
86	100	64	79	42	66	20	43
85	98	63	78	41	65	19	41
84	97	62	78	40	64	18	39
83	95	61	77	39	64	17	38
82	94	60	77	38	63	16	36
81	93	59	76	37	62	15	34
80	92	58	75	36	61	14	32
79	90	57	75	35	60	13	30
78	89	56	74	34	59	12	28
77	88	55	74	33	58	11	26
76	87	54	73	32	58	10	24
75	86	53	73	31	57	9	22
74	86	52	72	30	55	8	19
73	85	51	72	29	54	7	17
72	84	50	71	28	53	6	15
71	83	49	70	27	52	5	12
70	83	48	70	26	51	4	10
69	82	47	69	25	50	3	7
68	81	46	69	24	48	2	5
67	81	45	68	23	47	1	2
66	80	44	67	22	46	0	0
65	79	43	67	21	44		

Examination
June 2012
Geometry

GEOMETRY REFERENCE SHEET

Volume	Cylinder	$V = Bh$, where B is the area of the base
	Pyramid	$V = \frac{1}{3}Bh$, where B is the area of the base
	Right circular cone	$V = \frac{1}{3}Bh$, where B is the area of the base
	Sphere	$V = \frac{4}{3}\pi r^3$

Lateral area (L)	Right circular cylinder	$L = 2\pi rh$
	Right circular cone	$L = \pi r\ell$, where ℓ is the slant height

Surface area	Sphere	$SA = 4\pi r^2$

PART I

Answer all 28 questions in this part. Each correct answer will receive 2 credits. No partial credit will be allowed. For each question, write in the space provided the number preceding the word or expression that best completes the statement or answers the question.　[56 credits]

1 Triangle *ABC* is graphed on the set of axes below.

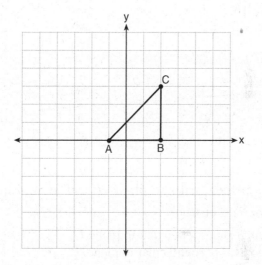

Which transformation produces an image that is similar to, but *not* congruent to, △*ABC*?

(1) $T_{2,3}$

(3) $r_{y\,=\,x}$

(2) D_2

(4) R_{90}

1 _____

2 A student wrote the sentence "4 is an odd integer." What is the negation of this sentence and the truth value of the negation?

(1) 3 is an odd integer; true
(2) 4 is not an odd integer; true
(3) 4 is not an even integer; false
(4) 4 is an even integer; false   2 _____

3 As shown in the diagram below, $\overleftrightarrow{EF}$ intersects planes $\mathcal{P}$, Q, and $\mathcal{R}$.

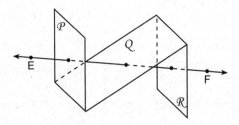

If $\overleftrightarrow{EF}$ is perpendicular to planes $\mathcal{P}$ and $\mathcal{R}$, which statement must be true?

(1) Plane $\mathcal{P}$ is perpendicular to plane Q.
(2) Plane $\mathcal{R}$ is perpendicular to plane $\mathcal{P}$.
(3) Plane $\mathcal{P}$ is parallel to plane Q.
(4) Plane $\mathcal{R}$ is parallel to plane $\mathcal{P}$.   3 _____

4 In the diagram below, *LATE* is an isosceles trapezoid with $\overline{LE} \cong \overline{AT}$, *LA* = 24, *ET* = 40, and *AT* = 10. Altitudes $\overline{LF}$ and $\overline{AG}$ are drawn.

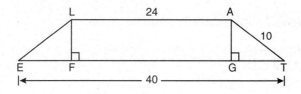

What is the length of $\overline{LF}$?

(1) 6   (3) 3

(2) 8   (4) 4   4 _____

5 In the diagram below of circle *O*, diameter $\overline{AB}$ is parallel to chord $\overline{CD}$.

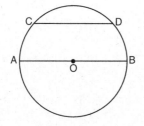

If m$\overarc{CD}$ = 70, what is m$\overarc{AC}$?

(1) 110   (3) 55

(2) 70   (4) 35   5 _____

6 In the diagram below of $\overline{ABCD}$, $\overline{AC} \cong \overline{BD}$.

Using this information, it could be proven that

(1) $BC = AB$
(2) $AB = CD$
(3) $AD - BC = CD$
(4) $AB + CD = AD$ 6 _____

7 The diameter of a sphere is 15 inches. What is the volume of the sphere, to the *nearest tenth of a cubic inch*?

(1) 706.9   (3) 2827.4
(2) 1767.1   (4) 14,137.2   7 _____

8 The diagram below shows the construction of $\overrightarrow{AB}$ through point P parallel to $\overleftrightarrow{CD}$.

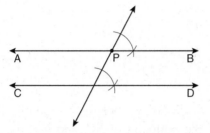

Which theorem justifies this method of construction?

(1) If two lines in a plane are perpendicular to a transversal at different points, then the lines are parallel.

(2) If two lines in a plane are cut by a transversal to form congruent corresponding angles, then the lines are parallel.

(3) If two lines in a plane are cut by a transversal to form congruent alternate interior angles, then the lines are parallel.

(4) If two lines in a plane are cut by a transversal to form congruent alternate exterior angles, then the lines are parallel.

8 _____

9 Parallelogram $ABCD$ has coordinates $A(1,5)$, $B(6,3)$, $C(3,-1)$, and $D(-2,1)$. What are the coordinates of E, the intersection of diagonals $\overline{AC}$ and $\overline{BD}$?

(1) (2,2)   (3) (3.5,2)

(2) (4.5,1)   (4) (-1,3)

9 _____

10 What is the equation of a circle whose center is 4 units above the origin in the coordinate plane and whose radius is 6?

(1) $x^2 + (y - 6)^2 = 16$
(2) $(x - 6)^2 + y^2 = 16$
(3) $x^2 + (y - 4)^2 = 36$
(4) $(x - 4)^2 + y^2 = 36$

10 _____

11 In the diagram of $\triangle ABC$ shown below, D is the midpoint of $\overline{AB}$, E is the midpoint of $\overline{BC}$, and F is the midpoint of $\overline{AC}$.

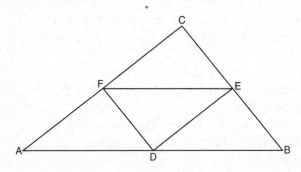

If $AB = 20$, $BC = 12$, and $AC = 16$, what is the perimeter of trapezoid $ABEF$?

(1) 24   (3) 40
(2) 36   (4) 44

11 _____

12 In the diagram below, △*LMO* is isosceles with *LO* = *MO*.

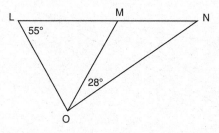

If m∠*L* = 55 and m∠*NOM* = 28, what is m∠*N*?

(1) 27   (3) 42

(2) 28   (4) 70   12 _____

13 If $\overleftrightarrow{AB}$ is contained in plane $\mathcal{P}$, and $\overleftrightarrow{AB}$ is perpendicular to plane $\mathcal{R}$, which statement is true?

(1) $\overleftrightarrow{AB}$ is parallel to plane $\mathcal{R}$.

(2) Plane $\mathcal{P}$ is parallel to plane $\mathcal{R}$.

(3) $\overleftrightarrow{AB}$ is perpendicular to plane $\mathcal{P}$.

(4) Plane $\mathcal{P}$ is perpendicular to plane $\mathcal{R}$.   13 _____

14 In the diagram below of $\triangle ABC$, $\overline{AE} \cong \overline{BE}$, $\overline{AF} \cong \overline{CF}$, and $\overline{CD} \cong \overline{BD}$.

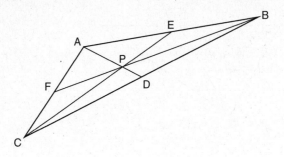

Point P must be the

(1) centroid   (3) incenter

(2) circumcenter   (4) orthocenter   14 _____

15 What is the equation of the line that passes through the point (–9,6) and is perpendicular to the line $y = 3x - 5$?

(1) $y = 3x + 21$ (3) $y = 3x + 33$

(2) $y = -\dfrac{1}{3}x - 3$ (4) $y = -\dfrac{1}{3}x + 3$ 15 _____

16 In the diagram of $\triangle ABC$ shown below, $\overline{DE} \parallel \overline{BC}$.

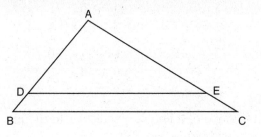

If $AB = 10$, $AD = 8$, and $AE = 12$, what is the length of $\overline{EC}$?

(1) 6　　　　　　　　　　(3) 3

(2) 2　　　　　　　　　　(4) 15　　　　　　　16 _____

17 What is the length of $\overline{AB}$ with endpoints $A(-1,0)$ and $B(4,-3)$?

(1) $\sqrt{6}$　　　　　　　　(3) $\sqrt{34}$

(2) $\sqrt{18}$　　　　　　　(4) $\sqrt{50}$　　　　17 _____

18 The sum of the interior angles of a polygon of n sides is

(1) 360

(3) $(n - 2) \cdot 180$

(2) $\dfrac{360}{n}$

(4) $\dfrac{(n - 2) \cdot 180}{n}$

18 _____

19 What is the slope of a line perpendicular to the line whose equation is $20x - 2y = 6$?

(1) -10

(3) 10

(2) $-\dfrac{1}{10}$

(4) $\dfrac{1}{10}$

19 _____

20 Which graph represents a circle whose equation is $(x + 2)^2 + y^2 = 16$?

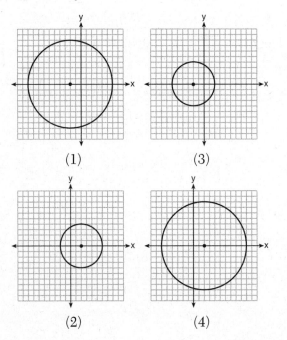

(1)   (3)

(2)   (4)

20 _____

21 In circle O shown below, diameter $\overline{DB}$ is perpendicular to chord $\overline{AC}$ at E.

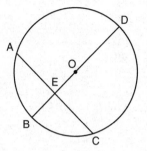

If $DB = 34$, $AC = 30$, and $DE > BE$, what is the length of $\overline{BE}$?

(1) 8   (3) 16

(2) 9   (4) 25   21 _____

22 In parallelogram $ABCD$ shown below, diagonals $\overline{AC}$ and $\overline{BD}$ intersect at E.

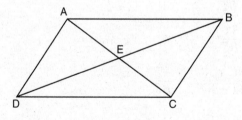

Which statement must be true?

(1) $\overline{AC} \cong \overline{DB}$ (3) $\triangle AED \cong \triangle CEB$

(2) $\angle ABD \cong \angle CBD$ (4) $\triangle DCE \cong \triangle BCE$ 22 _____

23 Which equation of a circle will have a graph that lies entirely in the first quadrant?

(1) $(x - 4)^2 + (y - 5)^2 = 9$

(2) $(x + 4)^2 + (y + 5)^2 = 9$

(3) $(x + 4)^2 + (y + 5)^2 = 25$

(4) $(x - 5)^2 + (y - 4)^2 = 25$ 23 ____

24 In the diagram below, $\triangle ABC \sim \triangle RST$.

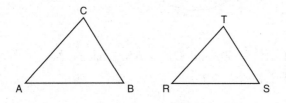

Which statement is *not* true?

(1) $\angle A \cong \angle R$

(2) $\dfrac{AB}{RS} = \dfrac{BC}{ST}$

(3) $\dfrac{AB}{BC} = \dfrac{ST}{RS}$

(4) $\dfrac{AB + BC + AC}{RS + ST + RT} = \dfrac{AB}{RS}$ 24 ____

25 In the diagram below of $\triangle ABC$, $\overline{BC}$ is extended to D.

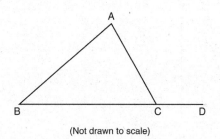

(Not drawn to scale)

If $m\angle A = x^2 - 6x$, $m\angle B = 2x - 3$, and $m\angle ACD = 9x + 27$, what is the value of x?

(1) 10   (3) 3

(2) 2   (4) 15   25 _____

26 An equation of the line that passes through $(2,-1)$ and is parallel to the line $2y + 3x = 8$ is

(1) $y = \dfrac{3}{2}x - 4$ (3) $y = -\dfrac{3}{2}x - 2$

(2) $y = \dfrac{3}{2}x + 4$ (4) $y = -\dfrac{3}{2}x + 2$ 26 _____

27 The graph below shows $\overline{JT}$ and its image, $\overline{J'T'}$, after a transformation.

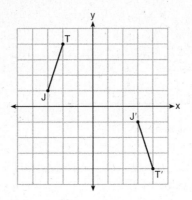

Which transformation would map $\overline{JT}$ onto $\overline{J'T'}$?

(1) translation
(2) glide reflection
(3) rotation centered at the origin
(4) reflection through the origin   27 _____

28 Which reason could be used to prove that a parallelogram is a rhombus?

(1) Diagonals are congruent.
(2) Opposite sides are parallel.
(3) Diagonals are perpendicular.
(4) Opposite angles are congruent.   28 _____

PART II

Answer all 6 questions in this part. Each correct answer will receive 2 credits. Clearly indicate the necessary steps, including appropriate formula substitutions, diagrams, graphs, charts, etc. For all questions in this part, a correct numerical answer with no work shown will receive only . credit. [12 credits]

29 Triangle *TAP* has coordinates $T(-1,4)$, $A(2,4)$, and $P(2,0)$.

On the set of axes below, graph and label $\triangle T'A'P'$, the image of $\triangle TAP$ after the translation $(x,y) \rightarrow (x-5,y-1)$.

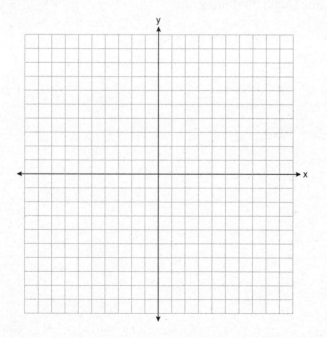

30 In the diagram below, $\ell \parallel m$ and $\overline{QR} \perp \overline{ST}$ at R.

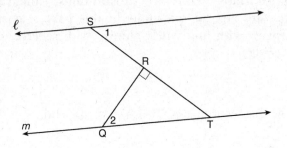

If $m\angle 1 = 63$, find $m\angle 2$.

31 Two lines are represented by the equations $x + 2y = 4$ and $4y - 2x = 12$. Determine whether these lines are parallel, perpendicular, or neither. Justify your answer.

32 Using a compass and straightedge, construct the bisector of ∠CBA.
[Leave all construction marks.]

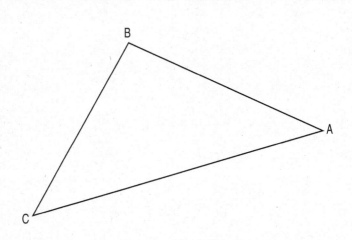

33 The cylindrical tank shown in the diagram below is to be painted. The tank is open at the top, and the bottom does *not* need to be painted. Only the outside needs to be painted. Each can of paint covers 600 square feet. How many cans of paint must be purchased to complete the job?

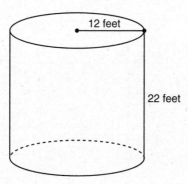

12 feet

22 feet

34 On the set of axes below, graph the locus of points that are 4 units from the line $x = 3$ and the locus of points that are 5 units from the point $(0,2)$. Label with an **X** *all* points that satisfy *both* conditions.

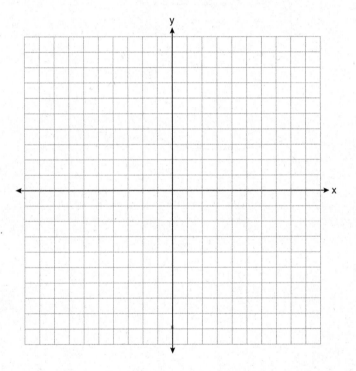

PART III

Answer all 3 questions in this part. Each correct answer will receive 4 credits. Clearly indicate the necessary steps, including appropriate formula substitutions, diagrams, graphs, charts, etc. For all questions in this part, a correct numerical answer with no work shown will receive only 1 credit. [12 credits]

35 Given: $\overline{AD}$ bisects $\overline{BC}$ at E.
$\overline{AB} \perp \overline{BC}$
$\overline{DC} \perp \overline{BC}$

Prove: $\overline{AB} \cong \overline{DC}$

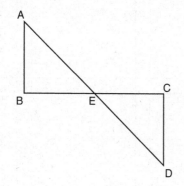

36 The coordinates of trapezoid *ABCD* are *A*(–4,5), *B*(1,5), *C*(1,2), and *D*(–6,2). Trapezoid *A″B″C″D″* is the image after the composition $r_{x\text{-axis}} \circ r_{y\,=\,x}$ is performed on trapezoid *ABCD*. State the coordinates of trapezoid *A″B″C″D″*.

[The use of the set of axes below is optional.]

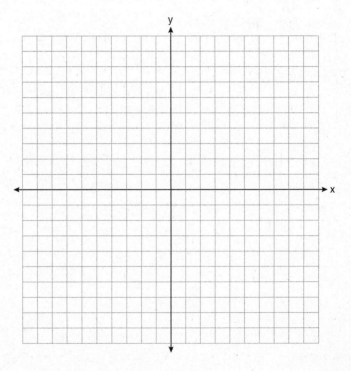

37 In the diagram below of circle O, chords $\overline{RT}$ and $\overline{QS}$ intersect at M. Secant $\overline{PTR}$ and tangent $\overline{PS}$ are drawn to circle O. The length of $\overline{RM}$ is two more than the length of $\overline{TM}$, $QM = 2$, $SM = 12$, and $PT = 8$.

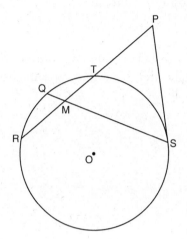

Find the length of $\overline{RT}$.

Find the length of $\overline{PS}$.

PART IV

Answer the question in this part. A correct answer will receive 6 credits. Clearly indicate the necessary steps, including appropriate formula substitutions, diagrams, graphs, charts, etc. A correct numerical answer with no work shown will receive only 1 credit.   [6 credits]

38 On the set of axes below, solve the system of equations graphically and state the coordinates of all points in the solution.

$$y = (x - 2)^2 - 3$$
$$2y + 16 = 4x$$

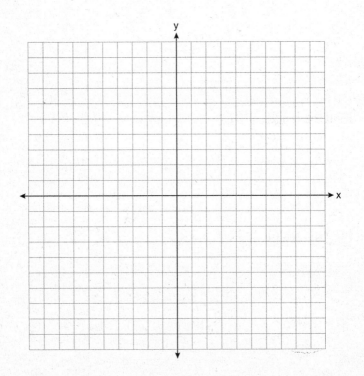

Answers
June 2012
Geometry

Answer Key

PART I

1. (2)	**8.** (2)	**15.** (4)	**22.** (3)
2. (2)	**9.** (1)	**16.** (3)	**23.** (1)
3. (4)	**10.** (3)	**17.** (3)	**24.** (3)
4. (1)	**11.** (4)	**18.** (3)	**25.** (4)
5. (3)	**12.** (1)	**19.** (2)	**26.** (4)
6. (2)	**13.** (4)	**20.** (3)	**27.** (2)
7. (2)	**14.** (1)	**21.** (2)	**28.** (3)

PART II

29. $T'(-6,3)$, $A'(-3,3)$, $P'(-3,-1)$

30. 27

31. Neither parallel or perpendicular

32. See the *Answers Explained* section.

33. 3

34. See the *Answers Explained* section.

PART III

35. See the *Answers Explained* section.

36. $A''(5,4)$, $B''(5,-1)$, $C''(2,-1)$, $D''(2,6)$

37. $RT = 10$, $PS = 12$

PART IV

38. $(3,-2)$

In **PARTS II–IV** you are required to show how you arrived at your answers. For sample methods of solutions, see the *Answers Explained* section.

Answers Explained

PART I

1. Triangle *ABC* is graphed on the set of axes below. You are asked to identify from among the answer choices the transformation that produces an image that is similar to, but *not* congruent to, △*ABC*.

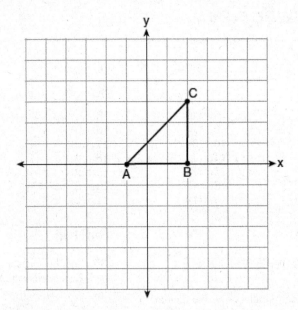

Translations, reflections, and rotations always produce images that are congruent to the original figure. A dilation with a scale factor other than 1 always produces an image that is similar to, but not congruent to, the original figure. If △*A'B'C'* is the image of △*ABC* under a dilation with a scale factor of 2, denoted by D_2, each side of △*A'B'C'* is two times the length of the corresponding side of △*ABC* so that △*A'B'C'* ~ △*ABC*.

The correct choice is (**2**).

2. The negation of a statement is the statement that has the opposite truth value, typically formed by inserting the word "not." The negation of the statement "4 is an odd integer" is the statement "4 is not an odd integer," which is true.

The correct choice is **(2)**.

3. As shown in the diagram below, $\overleftrightarrow{EF}$ intersects planes $\mathcal{P}$, $\mathcal{Q}$, and $\mathcal{R}$.

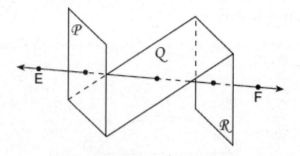

If two planes are perpendicular to the same line, they are parallel. Since it is given that $\overleftrightarrow{EF}$ is perpendicular to planes $\mathcal{P}$ and $\mathcal{R}$ these two planes must be parallel to each other.

The correct choice is **(4)**.

4. In the accompanying diagram of isosceles trapezoid $LATE$, $\overline{LE} \cong \overline{AT}$, $LA = 24$, $ET = 40$, and $AT = 10$. You are asked to find the length of altitude $\overline{LF}$.

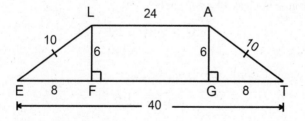

Since $LAGF$ is a rectangle, $FG = 24$:

$$EF + GT = 40 - 24$$
$$= 16$$

Since *LATE* is an isosceles trapezoid, *LE* = *AT* = 10. Because right triangle *EFL* is congruent to right triangle *TGA*:

$$EF = GT$$

$$= \frac{1}{2} \times 16$$

$$= 8$$

The lengths of the sides of right triangle *EFL* form a 6-8-10 Pythagorean triple where *LF* = 6. The length of $\overline{LF}$ is 6.

The correct choice is (**1**).

5. In the accompanying diagram of circle *O*, diameter $\overline{AB}$ is parallel to chord $\overline{CD}$, and $m\widehat{CD}$ = 70.

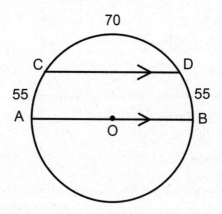

The degree measure of a semicircle is 180:

$$m\widehat{AC} + m\widehat{BD} = 180 - 70$$

$$= 110$$

Since parallel chords cut off congruent arcs:

$$m\widehat{AC} = m\widehat{BD}$$

$$= \frac{1}{2} \times 110$$

$$= 55$$

The correct choice is (**3**).

6. In the accompanying diagram of $\overline{ABCD}$, $AD \cong BD$.

Use this information to determine the following:

$$\overset{AC}{\overbrace{AB+BC}} = \overset{BD}{\overbrace{BC+CD}}$$
$$\underline{\;\;\;\;-BC = -BC\;\;\;\;}$$
$$AB \quad = \quad CD$$

The correct choice is **(2)**.

7. The volume, V, of a sphere is given by the formula, $V = \dfrac{4}{3}\pi r^3$, where r is the radius of the sphere. If the diameter of the given sphere is 15 inches, then its radius is 7.5 inches:

$$V = \frac{4}{3}\pi(7.5)^3$$
$$\approx 1767.145868 \text{ in}^3$$

The volume of the sphere to the *nearest tenth of a cubic inch* is 1767.1.

The correct choice is **(2)**.

8. The accompanying diagram shows a construction in which the angle formed by the transversal and $\overleftrightarrow{CD}$ is copied using P as its vertex. The construction thereby forms congruent corresponding angles, making $\overleftrightarrow{AB}$ and $\overleftrightarrow{CD}$ parallel.

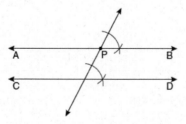

The correct choice is **(2)**.

9. Parallelogram $ABCD$ has coordinates $A(1,5)$, $B(6,3)$, $C(3,-1)$, and $D(-2,1)$. Since the diagonals of a parallelogram bisect each other, the coordinates of the point of intersection of diagonals $\overline{AC}$ and $\overline{BD}$ can be determined by finding the midpoint of either diagonal. The coordinates of the midpoint of diagonal $\overline{AC}$ are the averages of the corresponding x and y coordinates of its endpoints:

$$\text{midpoint } \overline{AC} = \left(\frac{1+3}{2}, \frac{5+(-1)}{2}\right)$$

$$= \left(\frac{4}{2}, \frac{4}{2}\right)$$

$$= (2,2)$$

The correct choice is **(1)**.

10. The center-radius form of the equation of a circle centered at (h,k) with a radius of r is $(x - h)^2 + (y - k)^2 = r^2$. If the center of a circle with radius 6 is 4 units above the origin in the coordinate plane, then $r = 6$ and $(h,k) = (0,4)$. So the equation of the circle is $(x - 0)^2 + (y - 4)^2 = 6^2$, which simplifies to $x^2 + (y - 4)^2 = 36$.

The correct choice is **(3)**.

11. In the accompanying diagram of $\triangle ABC$, D, E, and F are midpoints. In addition, $AB = 20$, $BC = 12$, and $AC = 16$.

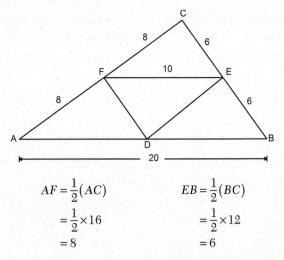

$$AF = \frac{1}{2}(AC) \qquad\qquad EB = \frac{1}{2}(BC)$$

$$= \frac{1}{2}\times 16 \qquad\qquad\quad = \frac{1}{2}\times 12$$

$$= 8 \qquad\qquad\qquad\quad = 6$$

Line segment *EF* connects the midpoints of two sides of a triangle. Therefore $\overline{EF}$ is parallel to the third side and one-half of its length:

$$EF = \frac{1}{2}(AB)$$
$$= \frac{1}{2} \times 20$$
$$= 10$$

The perimeter of trapezoid *ABEF* is *AF* + *AB* + *EB* + *EF*,

$$AF + AB + EB + EF = 8 + 20 + 6 + 10$$
$$= 44$$

The correct choice is (**4**).

12. In the accompanying diagram of isosceles $\triangle LMO$, *LO* = *MO*, m∠*L* = 55, and m∠*NOM* = 28. You are asked to find m∠*N*.

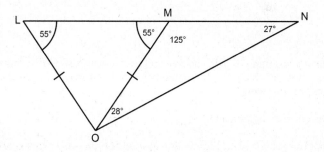

Since base angles of an isosceles triangle are congruent,

$$m\angle LMO = 55.$$

Since *LMO* and *NMO* are supplemental:

$$m\angle NMO = 180 - 55$$
$$= 125$$

The sum of the measures of the angles of a triangle is 180. In $\triangle OMN$:

$$m\angle N + 28 + 125 = 180$$
$$m\angle N + 153 = 180$$
$$m\angle N = 180 - 153$$
$$= 27$$

The correct choice is (**1**).

13. If line $\overleftrightarrow{AB}$ is perpendicular to plane $\mathcal{R}$, then any plane that contains $\overleftrightarrow{AB}$ must also be perpendicular to plane $\mathcal{R}$. Since it is given that $\overleftrightarrow{AB}$ is contained in plane $\mathcal{P}$, plane $\mathcal{P}$ is perpendicular to plane $\mathcal{R}$.

The correct choice is **(4)**.

14. In the accompanying diagram of $\triangle ABC$, $\overline{AE} \cong \overline{BE}$, $\overline{AF} \cong \overline{CF}$, and $\overline{CD} \cong \overline{BD}$.

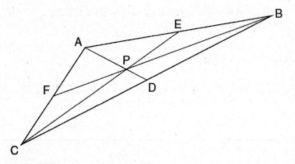

This means that $\overline{CE}$, $\overline{BF}$, and $\overline{AD}$ are the medians of $\triangle ABC$. Since point P is the point at which the three medians of the triangle intersect, it is the centroid of the triangle.

The correct choice is **(1)**.

15. You are asked to find an equation of the line that passes through the point $(-9,6)$ and is perpendicular to the line $y = 3x - 5$. Since the line $y = 3x - 5$ is written in $y = mx + b$ slope-intercept form, its slope is 3, which is the coefficient of the x-term.

Perpendicular lines have slopes that are negative reciprocals. So the slope of the required line is $-\frac{1}{3}$, the negative reciprocal of 3.

The equation of the required line has the form $y = -\frac{1}{3}x + b$. Since it is given that this line passes through $(-9,6)$, find b by substituting the coordinates of this point in $y = -\frac{1}{3}x + b$:

$$6 = -\frac{1}{3}(-9) + b$$
$$6 = 3 + b$$
$$b = 6 - 3$$
$$= 3$$

The equation of the required line is $y = -\frac{1}{3}x + 3$

The correct choice is **(4)**.

16. In the accompanying diagram of $\triangle ABC$, $\overline{DE} \parallel \overline{BC}$, $AB = 10$, $AD = 8$, and $AE = 12$. You are asked to find the length of $\overline{EC}$.

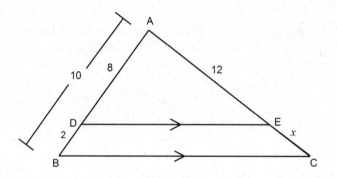

A line parallel to one side of a triangle and intersecting the other two sides divides these other two sides proportionally:

$$\frac{AD}{BD} = \frac{AE}{EC}$$
$$\frac{8}{10-8} = \frac{12}{x}$$
$$\frac{8}{2} = \frac{12}{x}$$
$$\frac{4}{1} = \frac{12}{x}$$
$$4x = 12$$
$$\frac{4x}{4} = \frac{12}{4}$$
$$x = 3$$

The correct choice is **(3)**.

17. To find the length of $\overline{AB}$ with endpoints $A(-1,0)$ and $B(4,-3)$, use the distance formula $AB = \sqrt{(x_2 - x_1)^2 + (y_2 - y_1)^2}$, where $(x_1, y_1) = A(-1,0)$ and $(x_2, y_2) = B(4,-3)$:

$$AB = \sqrt{(4-(-1))^2 + (-3-0)^2}$$
$$= \sqrt{(4+1)^2 + (-3)^2}$$
$$= \sqrt{25+9}$$
$$= \sqrt{34}$$

The correct choice is **(3)**.

18. If a polygon has n sides, then the polygon can be subdivided into $n-2$ nonoverlapping triangles as illustrated in the accompanying diagram for a polygon with 5 sides.

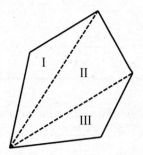

If $n = 5$, then $5 - 2 = 3$ triangles can be formed.

Since the sum of the measures of the interior angles of each triangle is 180, the sum of the measures of the interior angles of a polygon of n sides is $(n-2) \cdot 180$.

The correct choice is **(3)**.

19. To find the slope of a line perpendicular to the line whose equation is $20x - 2y = 6$, first solve the equation for y in terms of x:

$$20x - 2y = 6$$
$$-2y = -20x + 6$$
$$\frac{-2y}{-2} = \frac{-20x}{-2} + \frac{6}{2}$$
$$y = 10x + 3$$

Since the equation is now in $y = mx + b$ slope-intercept form, the slope of the given line is 10, the coefficient of the x-term. Perpendicular lines have slopes that are negative reciprocals. So the slope of a line perpendicular to the given line is $-\frac{1}{10}$, which is the negative reciprocal of 10.

The correct choice is **(2)**.

20. The center-radius form of the equation of a circle centered at (h,k) with a radius of r is $(x - h)^2 + (y - k)^2 = r^2$. The equation $(x + 2)^2 + y^2 = 16$ can be written in center-radius form as $(x - (-2))^2 + (y - 0)^2 = 4^2$. This equation means the circle's center is at $(-2,0)$ and the radius is 4, as shown in choice (3).

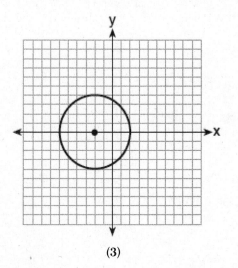

(3)

The correct choice is **(3)**.

21. In the accompanying diagram of circle O, diameter $\overline{DB}$ is perpendicular to chord $\overline{AC}$, $DB = 34$, $AC = 30$, and $DE > BE$. You are asked to find the length of $\overline{BE}$.

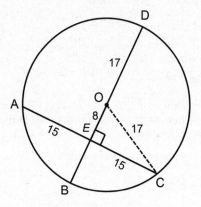

Draw radius OC. Since diameter $DB = 34$:

$$OC = \frac{1}{2} \times 34$$
$$= 17$$

A diameter that is perpendicular to a chord bisects the chord and its arcs. Since $AC = 30$:

$$AE = EC$$
$$= 15$$

In right triangle OEC:

$$(OE)^2 + (15)^2 = (17)^2$$
$$(OE)^2 = 289 - 225$$
$$OE = \sqrt{64}$$
$$= 8$$

Since radius $OB = 17$:

$$BE = 17 - 8$$
$$= 9$$

The length of $\overline{BE}$ is 9.

The correct choice is **(2)**.

22. In the accompanying diagram of parallelogram $ABCD$, diagonals $\overline{AC}$ and $\overline{BD}$ intersect at E. To determine which of the statements among the answer choices must be true, consider each statement in turn.

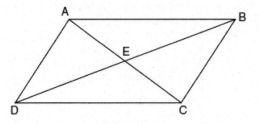

- Choice (1): $\overline{AC} \cong \overline{DB}$. Although the diagonals of a rectangle are congruent, the diagonals of a parallelogram are not necessarily congruent.   ✗

- Choice (2): $\angle ABD \cong \angle CBD$. Although the diagonals of a rhombus bisect its angles, the diagonals of a parallelogram do not necessarily bisect its angles.   ✗

- Choice (3): $\triangle AED \cong \angle CEB$. Since the diagonals of a parallelogram bisect each other and vertical angles are congruent, opposite pairs of triangles formed by the intersecting diagonals can be proved congruent by $SAS \cong SAS$. This statement must be true.   ✔

- Choice (4): $\triangle DCE \cong \angle BCE$ This statement is true only when the diagonals intersect at right angles as in a rhombus.   ✗

The correct choice is **(3)**.

23. To determine which answer choice includes the equation of a circle that lies entirely in the first quadrant, consider each equation in turn until you find the correct equation. Choice (1) gives the equation $(x - 4)^2 + (y - 5)^2 = 9$. The center of this circle is $(4,5)$ and the radius is 3. Since every point that is 3 units from $(4,5)$ lies in the first quadrant, the circle will lie entirely in the first quadrant as illustrated in the accompanying figure.

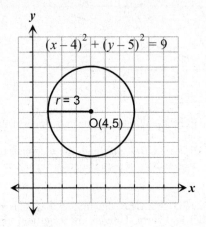

The correct choice is **(1)**.

24. In the accompanying diagram, $\triangle ABC \sim \triangle RST$.

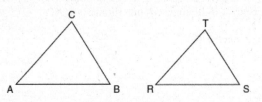

To determine the statement in the answer choices that is *not* true, consider each answer choice in turn.

- Choice (1): $\angle A \cong \angle R$. Since corresponding angles of similar triangles are congruent, this statement is true. ✔

- Choice (2): $\dfrac{AB}{RS} = \dfrac{BC}{ST}$. The lengths of corresponding sides of similar triangles are in proportion. Since $AB \leftrightarrow RS$ and $BC \leftrightarrow ST$, the proportion is true. ✔

- Choice (3): $\dfrac{AB}{BC} = \dfrac{ST}{RS}$. Since $\overline{AB}$ and $\overline{ST}$ are not corresponding sides, the proportion is not true. ✗

- Choice (4): $\dfrac{AB+BC+AC}{RS+ST+RT} = \dfrac{AB}{RS}$. Since the perimeters of similar triangles have the same ratio as the lengths of any pair of corresponding sides, the proportion is true. ✔

The correct choice is **(3)**.

25. In the accompanying diagram of $\triangle ABC$, $m\angle A = x^2 - 6x$, and $m\angle B = 2x - 3$, and $m\angle C = 9x + 27$.

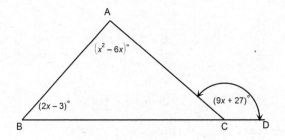

(Figure not drawn to scale)

Since the measure of an exterior angle of a triangle is equal to the sum of the measures of two nonadjacent interior angles:

$$\left(x^2 - 6x\right) + \left(2x - 3\right) = 9x + 27$$
$$x^2 - 4x - 3 = 9x + 27$$
$$x^2 - 4x - 9x - 3 - 27 = 0$$
$$x^2 - 13x - 30 = 0$$
$$(x - 15)(x + 2) = 0$$
$$x - 15 = 0 \quad \text{or} \quad x + 2 = 0$$
$$x = 15 \quad \text{or} \quad x = -2$$

One of the answer choices is 15, so this is the correct answer.

The correct choice is **(4)**

26. You must find an equation of the line that passes through the point $(2, -1)$ and is parallel to the line $2y + 3x = 8$.

Solve $2y + 3x = 8$ for y in terms of x, which gives $y = -\dfrac{3}{2}x + 4$. The equation is now in $y = mx + b$ slope-intercept form. The slope of the given line can be read from the equation as $-\dfrac{3}{2}$, which is the coefficient of the x-term.

Parallel lines have equal slopes. So the slope of the required line is $-\dfrac{3}{2}$.

The equation of the required line has the form $y = -\dfrac{3}{2}x + b$. Since it is given that this line passes through $(2, -1)$, find b by substituting the coordinates of this point in $y = -\dfrac{3}{2}x + b$:

$$-1 = -\frac{3}{\cancel{2}}\left(\cancel{2}\right) + b$$
$$-1 = -3 + b$$
$$-1 + 3 = b$$
$$2 = b$$

The equation of the required line is $y = -\dfrac{3}{2}x + 2$

The correct choice is **(4)**.

27. The accompanying graph shows $\overline{JT}$ and its image $\overline{J'T'}$ after a transformation. You must identify from among the answer choices the type of transformation that would map $\overline{JT}$ onto $\overline{J'T'}$.

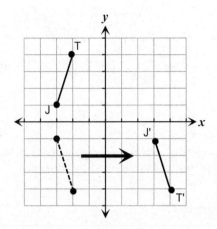

A glide reflection is the combination of a reflection in a line and a translation along that line. As shown in the accompanying diagram, reflecting $\overline{JT}$ over the x-axis and then translating the reflected image 6 units to the right maps $\overline{JT}$ onto $\overline{J'T'}$.

The correct choice is **(2)**.

28. You are asked to identity from among the answer choices the statement that could be used to prove that a parallelogram is a rhombus. Consider each answer choice in turn.

- Choice (1): Diagonals are congruent. This statement can be used to prove that a parallelogram is a rectangle but not that it is a rhombus. ✗

- Choice (2): Opposite sides are parallel. This is a property of all parallelograms, so it cannot be used to prove that a parallelogram is a rhombus. ✗

- Choice (3): Diagonals are perpendicular. If the diagonals of a parallelogram are perpendicular, the four triangles formed by the intersecting diagonals are congruent, as shown in the accompanying diagram. This means that all four sides of the parallelogram must be congruent.

Therefore the parallelogram is a rhombus. By SAS ≅ SAS, △I ≅ △II ≅ △III ≅ △IV. This means that $\overline{AB} \cong \overline{BC} \cong \overline{CD} \cong \overline{AD}$. ✔

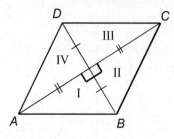

- Choice (4): Opposite angles are congruent. This is a property of all parallelograms, so it cannot be used to prove that a parallelogram is a rhombus. ✗

The correct choice is **(3)**.

PART II

29. It is given that triangle *TAP* has coordinates $T(-1,4)$, $A(2,4)$, and $P(2,0)$. On the set of axes provided in the test booklet, you need to graph and label $\triangle T'A'P'$, the image of $\triangle TAP$ after the translation $(x,y) \rightarrow (x - 5, y - 1)$. Determine the coordinates of $\triangle T'A'P'$:

$$T(-1,4) \rightarrow T'(-1-5, 4-1) = T''(-6,3)$$
$$A(2,4) \rightarrow A'(2-5, 4-1) = A'(-3,3)$$
$$P(2,0) \rightarrow P'(2-5, 0-1) = P'(-3,-1)$$

Graph $\triangle T'A'P'$ as shown on the accompanying set of axes.

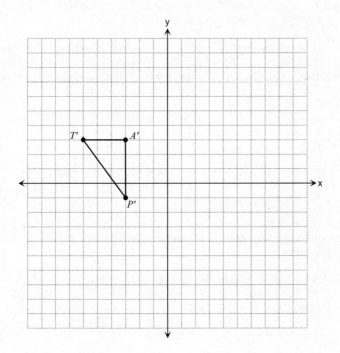

30. In the accompanying diagram, $\ell \parallel m$, $\overline{QR} \perp \overline{ST}$ at R, and m$\angle 1$ = 63.

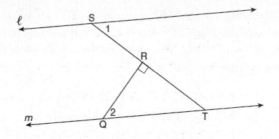

Parallel lines form congruent alternate interior angles. So m$\angle QTR$ = m$\angle 1$ = 63. Since the acute angles of a right triangle are complementary, m$\angle 2$ + 63 = 90. So m$\angle 2$ = 90 – 63 = **27**.

31. To determine whether the lines represented by the equations $x + 2y = 4$ and $4y - 2x = 12$ are parallel, perpendicular, or neither, write each of the equations in $y = mx + b$ slope-intercept form and then compare their slopes.

The equation $x + 2y = 4$ can be rewritten as $y = -\dfrac{1}{2}x + 2$. The slope of this line is $-\dfrac{1}{2}$.

The equation $4y - 2x = 12$ can be rewritten as $y = \dfrac{1}{2}x + 3$. The slope of this line is $\dfrac{1}{2}$.

The slopes of the lines are not equal, so the lines are not parallel. The slopes of the lines are not negative reciprocals, which means the lines are not perpendicular.

The lines are **neither parallel nor perpendicular.**

32. Using a compass and straightedge, you are asked to construct the bisector of ∠CBA.

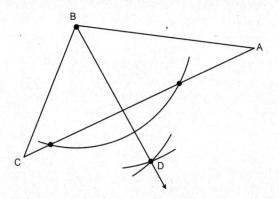

- Step 1: Using point B as a center and any convenient radius length, construct an arc that intersects $\overline{AC}$ at two different points, as shown in the accompanying figure.

- Step 2: Using each point of intersection as a center and the same radius length, construct a pair of arcs that intersect. Label the point at which the arcs intersect as D.

- Step 3: Use a straightedge to draw $\overline{BD}$.

$\overline{BD}$ bisects ∠CBA.

33. The cylindrical tank shown in the accompanying diagram is to be painted. The tank is open at the top, and the bottom does *not* need to be painted. Only the outside needs to be painted. If each can of paint covers 600 square feet, you need to determine the number of cans of paint that must be purchased to complete the job.

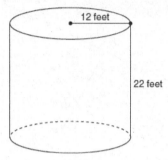

First determine the lateral surface area, L, of the cylinder using the formula $L = 2\pi rh$, where $r = 12$ and $h = 22$. The formula is given on the reference sheet that is located at the back of the acutal test booklet.

$$L = 2\pi \times 12 \times 22$$

$$\approx 1658.760921 \text{ ft}^2$$

Now determine the number of paint cans needed by dividing the surface area by 600:

$$\frac{1658.760921 \text{ ft}^2}{600 \text{ ft}^2/\text{can}} \approx 2.76 \text{ cans}$$

Hence, **3** cans of paint must be purchased.

34. On the set of axes provided in the test booklet, you are required to graph the locus of points that are 4 units from the line $x = 3$ and the locus of points that are 5 units from the point $(0,2)$. Label with an **X** *all* points that satisfy *both* conditions.

The locus of points that are 4 units from the vertical line $x = 3$ is a pair of vertical lines that are each 4 units from $x = 3$. The equations of these lines are $x = -1$ and $x = 7$, as shown in the accompanying diagram.

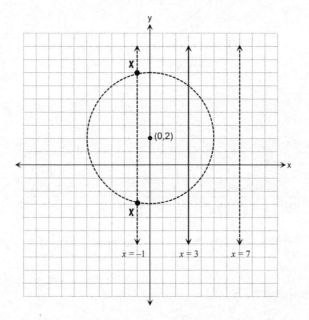

The locus of points that are 5 units from the point $(0,2)$ is a circle centered at $(0,2)$ with a radius of 5, as shown in the accompanying diagram.

The points where the graphs intersect represent the points that satisfy both locus conditions. Label with an **X** the two points at which the circle intersects the line $x = -1$, as indicated in the accompanying diagram.

PART III

35. In the accompanying diagram, $\overline{AD}$ bisects $\overline{BC}$ at E, $\overline{AB} \perp \overline{BC}$, and $\overline{DC} \perp \overline{BC}$. Prove that $\overline{AB} \cong \overline{DC}$.

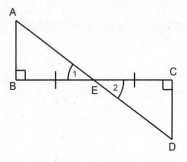

Plan: $\triangle ABE \cong \triangle DCE$ by ASA $\cong$ ASA

Statement	Reason
1. $\overline{AB} \perp \overline{BC}$ and $\overline{DC} \perp \overline{BC}$.	1. Given.
2. Angles B and C are right angles.	2. Perpendicular lines intersect to form right angles.
3. $\angle B \cong \angle C$. (Angle)	3. All right angles are congruent.
4. $\overline{AD}$ bisects $\overline{BC}$ at E.	4. Given.
5. $\overline{BE} \cong \overline{CE}$. (Side)	5. The midpoint of a line segment divides the segment into two congruent segments.
6. $\angle AEB \cong \angle DEC$. (Angle)	6. Vertical angles are congruent.
7. $\triangle ABE \cong \triangle DCE$.	7. ASA $\cong$ ASA.
8. $\overline{AB} \cong \overline{DC}$	8. Corresponding parts of congruent triangles are congruent.

36. The coordinates of trapezoid $ABCD$ are $A(-4,5)$, $B(1,5)$, $C(1,2)$, and $D(-6,2)$. Trapezoid $A''B''C''D''$ is the image after the composition $r_{x\text{-axis}} \circ r_{y=x}$ is performed on trapezoid $ABCD$. You must determine the coordinates of trapezoid $A''B''C''D''$.

The notation $r_{x\text{-axis}} \circ r_{y=x}$ represents a reflection over the line $y = x$ followed by a reflection of the image over the x-axis. In general, $r_{y=x}P(x,y) = P'(y,x)$ and $r_{x\text{-axis}}P(x,y) = P'(x,-y)$.

Reflect trapezoid $ABCD$ over the line $y = x$:

$$r_{y=x}A(-4,5) = A'(5,-4)$$

$$r_{y=x}B(1,5) = B'(5,1)$$

$$r_{y=x}C(1,2) = C'(2,1)$$

$$r_{y=x}D(-6,2) = D'(2,-6)$$

Reflect trapezoid $A'B'C'D'$ over the x-axis:

$$r_{x\text{-axis}}A'(5,-4) = A''(5,4)$$

$$r_{x\text{-axis}}B'(5,1) = B''(5,-1)$$

$$r_{x\text{-axis}}C'(2,1) = C''(2,-1)$$

$$r_{x\text{-axis}}D'(2,-6) = D''(2,6)$$

The coordinates of trapezoid $A''B''C''D''$ are **$A''(5,4)$, $B''(5,-1)$, $C''(2,-1)$, and $D''(2,6)$.**

37. In the accompanying diagram of circle O, chords $\overline{RT}$ and $\overline{QS}$ intersect at M. Secant $\overline{PTR}$ and tangent $\overline{PS}$ are drawn to circle O. The length of $\overline{RM}$ is two more than the length of $\overline{TM}$, $QM = 2$, $SM = 12$, and $PT = 8$. You are asked to find the lengths of $\overline{RT}$ and $\overline{PS}$.

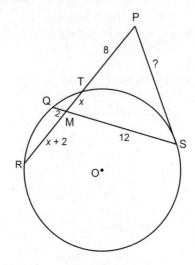

If x represents the length of $\overline{TM}$, then $x + 2$ represents the length of $\overline{RM}$. If two chords intersect inside a circle, the product of the segments of one chord is equal to the product of the segments of the other chord. Since chords $\overline{RT}$ and $\overline{PS}$ intersect inside circle O:

$$TM \times RM = QM \times SM$$
$$x(x+2) = 2 \times 12$$
$$x^2 + 2x = 24$$
$$x^2 + 2x - 24 = 0$$
$$(x-4)(x+6) = 0$$

$$x - 4 = 0 \quad \text{or} \quad x + 6 = 0$$
$$x = 4 \quad \text{or} \quad x = -6 \leftarrow \text{reject since } x \text{ must be positive}$$

Since $TM = x = 4$ and $RM = x + 2 = 4 + 2 = 6$:

$$RT = 4 + 6 = 10$$

If a secant ($\overline{PTR}$) and a tangent ($\overline{PS}$) are drawn to a circle from the same exterior point P, then the length of the tangent is the mean proportional between the lengths of the secant and its external segment ($\overline{PT}$):

$$\frac{PTR}{PS} = \frac{PS}{PT}$$

$$\frac{8+10}{PS} = \frac{PS}{8}$$

$$(PS)^2 = 8 \cdot 18$$

$$PS = 144$$

$$PS = \sqrt{144} = 12$$

The length of $\overline{RT}$ is **10**, and the length of ($\overline{PS}$) is **12**.

PART IV

38. You are given two equations:

$$y = (x - 2)^2 - 3$$

$$2y + 16 = 4x$$

To solve a linear-quadratic system of equations graphically, graph the equations on the same set of axes. Then find the coordinates of all points at which the graphs intersect.

- Step 1: Graph the quadratic equation. First find the vertex of the parabola. A parabola equation of the form $y = (x - h)^2 + k$ has its vertex at (h,k). For $y = (x - 2)^2 - 3$, $(h,k) = (2,-3)$ so the x-coordinate of the vertex of the parabola is 2.

- Step 2: Use your graphing calculator to create a table of values that is centered at $x = 2$:

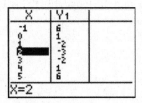

- Step 3: Now draw the parabola. Choose a suitable scale. Plot $(-1,6)$, $(0,1)$, $(1,-2)$, $(2,-3)$, $(3,-2)$ $(4,1)$, and $(5,6)$. Then connect these points with a smooth, U-shaped curve as shown in the accompanying figure. Label the graph with its equation.

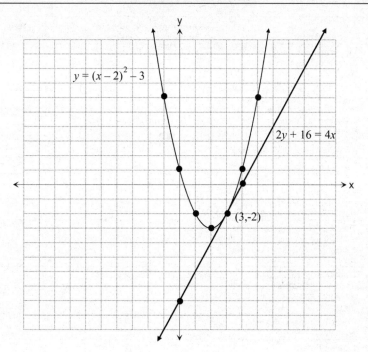

$y = (x - 2)^2 - 3$

$2y + 16 = 4x$

$(3,-2)$

- Step 4: Graph the linear equation on the same set of axes. Graph $2y + 16 = 4x$ by finding two convenient points on the line. If $x = 0$, $2y = -16$. So $y = \dfrac{-16}{2} = -8$. The line contains the point $(0,-8)$. If $y = 0$, $16 = 4x$. So $x = \dfrac{16}{4} = 4$. The line contains the point $(4,0)$. Plot these two points, and then draw a line through them as shown in the accompanying figure. Label the graph with its equation.

- Step 5: State the coordinates of all solution points. The graphs intersect at $(3,-2)$.

The solution is **(3,-2)** since it is the only point that satisfies both equations.

Topic	Question Numbers	Number of Points	Your Points	Your Percentage
1. Logic (negation, conjunction, disjunction, related conditionals, biconditional, logically equivalent statements, logical inference)	2	2		
2. Angle & Line Relationships (vertical angles, midpoint, altitude, median, bisector, supplementary angles, complementary angles, parallel and perpendicular lines)	6, 30	2 + 2 = 4		
3. Parallel & Perpendicular Planes	3, 13	2 + 2 = 4		
4. Angles of a Triangle & Polygon (sum of angles, exterior angle, base angles theorem, angles of a regular polygon)	12, 18, 25	2 + 2 + 2 = 6		
5. Triangle Inequalities (side length restrictions, unequal angles opposite unequal sides, exterior angle)	—	—		
6. Trapezoids & Parallelograms (includes properties of rectangle, rhombus, square, and isosceles trapezoid)	4, 22, 28	2 + 2 + 2 = 6		
7. Proofs Involving Congruent Triangles	35,	4		
8. Indirect Proof & Mathematical Reasoning	—	—		
9. Ratio & Proportion (includes similar polygons, proportions, proofs involving similar triangles, comparing linear dimensions, and areas of similar triangles)	16, 24	2 + 2 = 4		
10. Proportions formed by altitude to hypotenuse of right triangle; Pythagorean theorem	—	—		
11. Coordinate Geometry (slopes of parallel and perpendicular lines; slope-intercept and point-slope equations of a line; applications requiring midpoint, distance, or slope formulas; coordinate proofs)	9, 15, 17, 19, 26, 31	2 + 2 + 2 + 2 + 2 + 2 = 12		
12. Solving a Linear-Quadratic System of Equations Graphically	38	6		
13. Transformation Geometry (includes isometries, dilation, symmetry, composite transformations, transformations using coordinates)	1, 27, 29, 36	2 + 2 + 2 + 4 = 10		
14. Locus & Constructions (simple and compound locus, locus using coordinates, compass constructions)	8, 32, 34	2 + 2 + 2 = 6		

Topic	Question Numbers	Number of Points	Your Points	Your Percentage
15. Midpoint and Concurrency Theorems (includes joining midpoints of two sides of a triangle, three sides of a triangle, the sides of a quadrilateral, median of a trapezoid; centroid, orthocenter, incenter, and circumcenter)	11, 14	2 + 2 = 4		
16. Circles and Angle Measurement (includes ≅ tangents from same exterior point, radius ⊥ tangent, tangent circles, common tangents, arcs and chords, diameter ⊥ chord, arc length, area of a sector, center-radius equation, applying transformations)	5, 10, 20, 23	2 + 2 + 2 + 2 = 8		
17. Circles and Similar Triangles (includes proofs, segments of intersecting chords, tangent and secant segments)	21, 37	2 + 4 = 6		
18. Area of Plane Figures (includes area of a regular polygon)	—	—		
19. Measurement of Solids (volume and lateral area; surface area of a sphere; great circle; comparing similar solids)	7, 33	2 + 2 = 4		

Map to Core Curriculum

Content Band	Item Numbers
Geometric Relationships	3, 7, 13, 33
Constructions	8, 32
Locus	14, 34
Informal and Formal Proofs	2, 4, 5, 6, 11, 12, 16, 18, 21, 22, 24, 25, 28, 30, 35, 37
Transformational Geometry	1, 27, 29, 36
Coordinate Geometry	9, 10, 15, 17, 19, 20, 23, 26, 31, 38

How to Convert Your Raw Score to Your Geometry Regents Examination Score

The conversion chart on the following page must be used to determine your final score on the June 2012 Regents Examination in Geometry. To find your final exam score, locate in the column labeled "Raw Score" the total number of points you scored out of a possible 86 points. Since partial credit is allowed in Parts II, III, and IV of the test, you may need to approximate the credit you would receive for a solution that is not completely correct. Then locate in the adjacent column to the right the scale score that corresponds to your raw score. The scale score is your final Geometry Regents Examination score.

Regents Examination in Geometry—June 2012

Chart for Converting Total Test Raw Scores to Final Examination Scores (Scale Scores)

Raw Score	Scale Score	Raw Score	Scale Score	Raw Score	Scale Score	Raw Score	Scale Score
86	100	64	80	42	66	20	41
85	98	63	80	41	65	19	39
84	97	62	79	40	64	18	38
83	95	61	78	39	63	17	36
82	94	60	78	38	62	16	34
81	93	59	77	37	62	15	33
80	92	58	77	36	61	14	31
79	91	57	76	35	60	13	29
78	90	56	75	34	59	12	27
77	89	55	75	33	58	11	25
76	88	54	74	32	56	10	23
75	88	53	74	31	55	9	21
74	87	52	73	30	54	8	19
73	86	51	72	29	53	7	17
72	86	50	72	28	52	6	14
71	85	49	71	27	51	5	12
70	84	48	70	26	49	4	10
69	83	47	70	25	48	3	8
68	83	46	69	24	47	2	5
67	82	45	68	23	45	1	2
66	81	44	68	22	44	0	0
65	81	43	67	21	42		

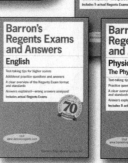

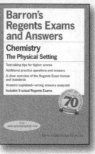
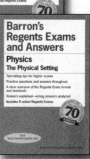

NOTES

NOTES

NOTES